D1528323

The Social Dynamics of Technology

Practice, Politics, and World Views

EDITED BY MARCIA-ANNE DOBRES
AND CHRISTOPHER R. HOFFMAN

SMITHSONIAN INSTITUTION PRESS
WASHINGTON AND LONDON

Copy editor: Vicky Macintyre
Production editor: Ruth G. Thomson
Designer: Linda McKnight

Library of Congress Cataloging-in-Publication Data

The social dynamics of technology : practice, politics, and world views / edited by Marcia-Anne Dobres and Christopher R. Hoffman.
p. cm.
Includes bibliographical references and index.
ISBN 1-56098-909-2 (alk. paper)
1. Material culture. 2. Technology—Sociological aspects.
3. Technology—Social aspects. I. Dobres, Marcia-Anne. II. Hoffman, Christopher R.
GN406.S6 1999
306.4′6—dc21 98-38120

British Library Cataloguing-in-Publication Data available

Manufactured in the United States of America
06 05 04 03 02 01 00 99 5 4 3 2 1

♾ The paper used in this publication meets the minimum requirements of the American National Standard for Information Sciences—Permanence of Paper for Printed Library Materials ANSI Z39.48-1984.

Contents

List of Illustrations

FIGURES

TABLE

Foreword

TIM INGOLD

ARCHAEOLOGISTS HAVE ALWAYS STUDIED ARTIFACTS, and the pages of anthropological monographs are full of details about the practices by which people secure a livelihood in this or that environment. Yet only recently have archaeology and anthropology come together around an explicit focus that has long remained on the sidelines: technology. What does it *mean* to make technology the center of theoretical concern? As yet, the answer to this question is far from clear, even among the advocates of an archaeology or anthropology of technology, some of whom are represented in this volume. Ever since the theoretical barriers that traditionally excluded technology from the domains of culture and social life came down, there has been nothing to prevent the concept of technology from spilling over from a narrow focus on tools and techniques to embrace the entire field of human endeavor. Is there anything, the skeptic might ask, about human culture and social life that is *not* technological? If not, what need is there for the concept of technology at all? Beyond stating the obvious, how does it help to know that *everything* is technological?

The first point to note in tackling this kind of definitional issue is that technology is not one of those concepts invented by anthropologists or archaeologists to assist in the organization and analysis of their material. Rather, it has come into general use as part of the movement of Western thought that also established humankind as an object of scientific inquiry. The same applies, of course, to such concepts as culture, society, humanity, and nature. Of the concept of society, it has been said that to use it is not to denote a thing, but to make a claim. Similarly, if we want to know what technology means, it will be of no avail to seek some delimited set of objects and activities in the world—as though we could point to them and say, "There, that's technology for you." Instead, we should be trying to find out what sorts of claims are being made with this word, and whether these claims are justified. And we must be prepared to face the possibility that they might *not* be justified, at least in terms of the understandings of the people whose works and activities are at issue. If, for example, these people have no word that corresponds to our "technology," this does not necessarily imply that there is a gap in their lexicon (perhaps because they have no cause to generalize about their practices and products), that could easily be filled by introducing an appropriate term. Rather, the claims that we advance with the concept of technology might simply make no sense to them.

A second point to recognize is that the meaning of "technology," like those of all words, is not fixed but has changed significantly in the course of history. One particularly noteworthy change concerns the relationship between technology and art. Nowadays, we tend to treat art and technology as referring to fields of endeavor that are in certain respects antithetical. This opposition, however, is a relatively recent development, going back—like the word "technology" itself—no further than the seventeenth century. "Technology" is derived from the Greek root *tekhnē,* and "art" from the Old French *ars,* but both *ars* and *tekhnē* once meant essentially the same thing, namely *skill* of the kind associated with craftsmanship. This is the sense preserved, on the one hand, in the concept of "technique" (that this term is borrowed from the French is significant, as I show below). On the other hand, it is preserved in terms such as "artful," "artifice," "artisan," and—of course—"artifact." Thus there appears to have been a historical progression from a traditional congruence of *ars* and *tekhnē,* around the notion of skill, to the modern opposition between art and technology. There is little doubt that this opposition is but one instance of the more general dichotomy in Western thought between freedom and necessity. That is why technology tends to be associated with the mechanical replication of the given rather than the creative production of novelty, and hence with what is objective and determined rather than what is subjective and spontaneous.

A third point to make about the concept of technology concerns its connection with what is commonly called "material culture." Why does the idea of "technology" invariably lead us to think in the first place of objects rather than practices, and of artifacts or tools rather than the processes into which they are incorporated? This is certainly one of the reasons why many anthropologists have been so dismissive of technology in the past, as they have of material culture in general. Their complaint is that an anthropology of technology deals only with lifeless objects, as presented in a museum, rather than with real-life processes as witnessed by anthropologists in the field. To focus on technology rather than social relations, it is claimed, is to leave out the people. Indeed, it is on precisely these grounds that anthropologists have often sought to distinguish their mode of inquiry from that of archaeology. The problem for archaeologists, it appears, is that they are always too late: by the time they arrive on the scene, the people have long been dead, and nothing remains but the things they used, or that provided the material scaffolding for their activity. Even if the people are not yet dead, the (ethno)archaeologist sometimes proceeds as if they were!

It seems that the study of technology has been compromised by its alignment on one side of the boundary between people and things. This volume provides ample proof of the artificiality of this boundary, and of the need to situate objects in the performative contexts of their use by skilled human agents if their meaning is to be properly understood. To study people and things *in their mutual relations,* as instantiated in skilled performance, is naturally to bring the interests of archaeology and anthropology together.

The assumption that technology consists essentially in systems of relations between things, as distinct from relations between persons, probably accounts for its rigid exclusion, until recently, from social and cultural theory. In the past, our theoretical mentors repeatedly warned us never to confuse technical relations with social relations. People guilty of such confusion were accused of vulgar materialism, or worse. Although there were disagreements as to whether technology had a positive impact on the form of culture or exerted no significant impact at all, all sides agreed that, whether prescriptive or permissive, technology was external to culture and society. Whatever its impact, technology could be left to technologists, just as climate could be left to climatologists and ecology to ecologists. It is precisely this *separation* of society and technology that many of the contributors to this volume set out to challenge. In doing so they apply much the same kind of critical strategy to the concept of technology that, two or three decades ago, anthropologists of the "substantive" school applied to the concept of economy. These anthropologists showed how "economy" and "society" became institutionally separate in the history of Western capitalism; how the "economic" is itself a product of this history; how in precapitalist societies economic relations are embedded in social relations; and how, with the development of market-oriented capitalism, economic life was progressively disembedded from social life.

Like economy and society, technology and society became institutionally separate in the recent history of the West, alongside the emergence of what could be called a "machine-theoretical" cosmology. We are only now becoming aware of the extent to which our theoretical concepts and suppositions are grounded in this institutional foundation. By the same token, we are just beginning to recognize the distortions and misunderstandings that can arise when concepts of recent Western provenance are retrojected into history and prehistory, or applied to the analysis of practices in non-Western societies. Three sources of misunderstanding, in particular, merit attention: what I call the "contemplative stance," the logocentrism inherent in the concept of technology, and the dichotomy between mechanics and semiotics.

The contemplative stance is the point of view of the spectator who, from an imagined position outside the world, regards the products of human activity as objects of value in themselves, detached from the contexts and relationships in which they came into being and are brought into play. It is this stance, as I indicated earlier, that leads to the identification of technology with material culture, that is, with things rather than the processes of both making and using. The labor process is, as it were, swallowed up by its products, work by "works." Yet the producers themselves may have little interest in the products, which, from their point of view, may be incidental to what really counts, namely the skilled, socially situated performances of making and using. For it is such performances that create identities and relationships. Skill, after all, is a property of people rather than of things, or more strictly, of people in their *relations* to the things around them.

However, "technology," formed as it is from a compounding of the roots *tekhnē* and *logos,* implies that skilled practice can, in some sense, be represented as the operation of a formal system of rules and principles. Wherein, one might ask, do these rules and principles reside? Here there is an important difference between French and Anglo-American understandings of the meaning of technology. The French technologist is someone engaged in the study of the skilled practices, or techniques, of everyday life. Technology is his or her discipline, and its principles comprise the framework of concepts and procedures that govern the inquiry. In Anglo-American usage, by contrast, the principles of technology are understood to be installed at the heart of the apparatus of production itself, whence they generate practice just as a program generates an output. Significantly, the concept of biology, also a French invention, has undergone a similar transformation in Anglo-American usage: from a set of principles for the study of living organisms to a set of (genotypic) programs installed in the organisms themselves and underwriting their development. It is no wonder, then, that Anglo-American authors have turned to French for a concept, namely "technique," to signal the kind of skilled practices that resist codification in terms of any formal system of rules and principles. Nor is it surprising that French scholarship has a long tradition of studying such skills that has no counterpart in Britain or North America.

Finally, I come to the distinction between mechanics and semiotics. Of a work of art, we are inclined to ask, "What does it mean (or represent)?" Of an item of technology, by contrast, we are more likely to ask, "How does it work?" An object that does not appear to work in any obvious way is assumed to be symbolic; conversely, if it does not seem to have any symbolic meaning, we search for a mechanical function. Put differently, to the extent that its form is determined by its function, an object is said to serve a technical purpose, but to the extent that its form is not so determined—and is in that respect "ornamental" or purely a matter of "style"—its purpose is supposed to be to convey some kind of social meaning. These distinctions, however, are exceedingly problematic. They are still commonly made, even by those archaeologists and anthropologists who have drawn explicit attention to the social dimensions of technological practice. Technology, they say, does not just produce physical effects in the world; it also serves to communicate or give expression to diverse cultural messages. However, this division between the communication of information and the alteration of the physical world rests on the supposition that human practitioners inhabit worlds of intersubjective meaning, or are caught up in "webs of significance," over and above the level of their material interventions. In short, it implies an initial separation of mind from world, or of "emic" from "etic" levels of reality.

Such separation is challenged by a phenomenological approach that would place the practitioner, right from the start, in an active perceptual engagement with the components of his or her environment. Advocates of this latter approach, among whom I number myself, would argue that social or cultural meaning, far from being "attached" to action, as signified to signifier, is rather immanent in the relationships between people and their surroundings, relationships that are a precondition for skilled performance. Views of the world are taken *in* it, they are not representations *of* it; they are ways of seeing, not pictures in the mind. As such they are continually generated in, rather than expressed by, the practices in which people are engaged. I believe that one of the reasons why anthropologists and archaeologists have had such difficulty in coming to grips with the notion of skill is that they remain more or less wedded to a cognitivist paradigm, which, by regarding technical action as the mechanical implementation of preconceived design, effectively forces a division between knowledge and practice. But the essence of skill surely lies in that knowledgeability inherent in the practitioner's own bodily movements in the world. To separate knowledge from practice is both to transform practical "know-how" into intellectual "know-that," and to reduce practice from creative doing or making to "merely mechanical" execution.

The problem of understanding the nature of skill, like other issues presented in this volume, depends for its resolution on the outcome of much wider debates in archaeology and anthropology. Since the convergence of these disciplines around the study of technology is in its infancy, ideas have still to coalesce into clearly identifiable positions. The perspectives offered here are as diverse as they are conflicting. What they all share, however, is a concern to bring technology back to life by reinserting it into the current of human activity and social relations, within which—and only within which—it can be effective. Whither this concern will eventually lead is too early to say. No one, however, can read this volume and still maintain, as was confidently asserted in the past, that technology and society are mutually exclusive domains of inquiry that should never be confused. This, in itself, is an index of how far we have come in the space of just a few years.

Acknowledgments

NO BOOK, EDITED OR OTHERWISE, becomes a product on its own. Its material production is a sequential process in which many people, with various sorts of knowledge, levels of expertise, ideas, and both personal and group agendas come together to forge a product that is not necessarily what any individual may have intended. This book stems from a series of conversations we began in 1992 at the University of California, Berkeley, while sharing the stress of dissertating. Our sincere gratitude to Meg Conkey and Ruth Tringham, who first brought us together, then provided an environment in which we could begin to imagine an alternative vision of prehistoric technology. Early on, our ideas were stimulated by Tim Ingold and Heather Lechtman, who served as a cheering squad and graciously shared their unpublished manuscripts and critical comments with us. From this fitful start, our project began to materialize thanks to Mike Schiffer, who agreed to publish "Social Agency and the Dynamics of Prehistoric Technology" in the *Journal of Archaeological Method and Theory* (1994), despite his own antipathy to many of its ideas. Positive reaction to that article motivated us to organize two symposia, which became the fodder for this collection: at the 1994 meetings of the American Anthropological Association in Atlanta, and the 1995 meetings of the Society for American Archaeology in Minneapolis. Participants at both gatherings did much to develop our fledgling perspective and take it in directions neither of us had envisioned, but from which it has certainly prospered. Among those not represented here, we thank Cathy Costin, Randy White, and Rita Wright. Our appreciation extends to Smithsonian Institution Press, to Mark Hirsch and Kate Gibbs, in particular, for recognizing the import of this topic, assembling a terrific production team, and moving the project steadily forward. Along the way, S. Terry Childs, Valentine Roux, Rita Wright, Matthew Johnson, and two anonymous reviewers provided astute advice, and David Wheeler provided technical assistance. Final thanks are due to the contributors to this volume, for bearing up under our editorial demands, for completing their tasks with diligence, and especially for sharing their work and ideas so as to further develop a human-centered vision of technology, past and present.

Contributors

S. Terry Childs
Archeology and Ethnology Program
National Park Service, Washington, D.C.

Marcia-Anne Dobres
Archaeological Research Facility
Department of Anthropology, University of California, Berkeley

Christopher R. Hoffman
Archaeological Research Facility
Department of Anthropology, University of California, Berkeley

Tim Ingold
Department of Social Anthropology
University of Manchester, United Kingdom

Roy Larick
Department of Anthropology
University of Massachusetts at Amherst

Heather Lechtman
Center for Materials Research in Archaeology and Ethnology
Massachusetts Institute of Technology

Pierre Matarasso
Centre National de la Recherche Scientifique (CNRS)
Centre International de Recherches sur l'Environement
et la Développement (CIRED)
URA 940, Montrouge, France

Bryan Pfaffenberger
Division of Technology, Culture, and Communication
School of Engineering and Applied Science
University of Virginia

Robin Ridington
Department of Anthropology (Emeritus)
University of British Columbia, Canada

Valentine Roux
Centre National de la Recherche Scientifique
Maison de l'Archéologie et de l'Ethnologie, Nanterre, France

Thomas Wake
Zooarchaeology Laboratory, The Institute of Archaeology
University of California, Los Angeles

1. Introduction: A Context for the Present and Future of Technology Studies

MARCIA-ANNE DOBRES AND
CHRISTOPHER R. HOFFMAN

> It is the signal task of the history of technology to decode and explain this culture embedded in material. (Pursell 1985:122)

HOW IS TECHNOLOGY IMPLICATED in expressing, reproducing, and contesting everyday social practices and culture-bound beliefs? In promoting political agency? In making and materializing culture? What makes technological practice at one and the same time a materially grounded social activity and a political statement (and even an ideology)? How are the tensions between different "spheres" of social life, such as economics, politics, social relations of production, and beliefs negotiated through technical practices engaged in at the scale of the everyday? These questions motivate this collection of essays on technology by sociocultural anthropologists and archaeologists. Although they echo an earlier generation of questions, they are recast here in light of theoretical and methodological trends now taking place across the social sciences.

As one of the most politically and symbolically charged of all human activities, technology is central both to human existence and to the very way human beings experience and make sense of their world. Yet, until quite recently, theories and methodologies for the study of technology rested on mechanistic perspectives that have all but denied a significant role to technical agency and shared belief systems. Often referred to as the Standard View (Pfaffenberger 1992:493–494), this decidedly modernist and instrumentalist view is now being scrutinized and challenged on many fronts (overview in Dobres and Hoffman 1994; Dobres forthcoming). In its stead, an alternative literature has begun to emerge that questions traditional assumptions about the relationship "between" technology and society by turning explicit attention to the sociopolitical dynamics and belief systems underlying technological practice through time and space. What is both remarkable and exciting is that this movement bridges sociocultural anthropology, archaeology, and the related disciplines of philosophy, history, sociology, and culture studies. Although numerous case studies now demonstrate the vast potential of this newer perspective, its theoretical and methodological contours are just beginning to come into focus (Dobres and Hoffman 1994:211–212). These recent trends are welcome and critically important additions to orthodox research on material culture production and use. In this volume, we explore some of their implications for a developing anthropology of technology.

Our intent in bringing together the particular studies in this volume is to highlight and discuss more fully three topics that we find are of considerable import in this emerging literature: (1) the dialectic of *technological practice* and social relations of production and their combined impact on the production, use, commodification, and exchange of end-products; (2) the *political nature* of technique and product in relation to *social agency* and identity; and (3) the structural importance of historically specific cultural systems of representation, or *world views,* in shaping the physical contours of technological practice, past and present. Through the case studies presented here we hope to convince the reader that technological practice, politics, and world views are inseparable facets of a universal and age-old human activity, one that no longer can be defined on the basis of materiality and functionality, or understood primarily as an economic pragmatic or rational *logos.* The following chapters therefore do not so much reflect the current state of technology studies as they advance an explicitly human-centered understanding of the dynamics of technological practice across time, space, and medium.

The topics on which this book concentrates—technological practice, politics, and world views—are left purposefully broad. The view we promote is that technology is a pervasive and powerful complex of mutually reinforcing sociomaterial practices structured by self- and group-interests, expressions of agency, identity and affiliation, cultural ways of comprehending and acting on the world, practical and esoteric knowledge, symbolic representations, and skill. These dynamics come together to create meaningful arenas in which humans simultaneously engage with each other and with their material world. Like Hughes (1979), we consider technology to be a web of social and material dynamics that together contribute to the making and remaking of society. While this may be an exceptionally broad position, especially for archaeologists whose primary data are material remains, we believe it best captures the essential nature of multiple and overlapping phenomena constituting what is too often conceptually limited to physical making and use.

Rather than privileging one over the others, our constructivist view stresses the dialectic of material, social, and symbolic factors in technological practice. Technology defined as a web of tangible and intangible dimensions is a position freely borrowed from philosophy and history (particularly Hughes 1979; Winner 1986a; overview in Dobres and Hoffman 1994:226–230) and subsequently modified. To weave cultural practices, beliefs, social relations, politics, and material realities into the very essence of technology is to complicate things considerably, especially for archaeologists still reeling under the weight of Hawkes's Ladder of Inference (1954). On the other hand, to conceptually limit our definition of technology to tangible, utilitarian, and practical object matter because it is seemingly more knowable is to cut off from that consideration an understanding of the dynamic subject matter that makes human technologies both possible and meaningful. That subject matter, of course, is people, their social relationships in productive endeavors, and their (not necessarily uniform) attitudes, values, and rules about making and using the material world.

For all its material grounding, our view of technology stresses the dynamic, ongoing, and socially constituted nature of sociotechnical activities. In this sense, we prefer to think of technology as a verb of action and interaction, rather than a noun of possession (Dobres 1995a, forthcoming). Because technology is an ever unfolding *process*, a "becoming," as it were, it necessarily interweaves the experiential making and use of material culture with the making and remaking of culture, and both with the making of social agents (Heidegger 1977; Ingold 1993; in this volume, Hoffman chapter 5, Dobres chapter 6, and Pfaffenberger chapter 7). One of the subtexts here is an explicit desire to counterbalance the century-old mechanistic view of technology pervading not only anthropological and archaeological studies but also folk notions (Pfaffenberger 1992). As the following chapters show, for different times, places, social formations, and material products, technology is caught in contextualized webs of social relations and divisions of labor, in the reciprocal ways that end-products define and structure social interaction and feed back on cultural values, and in the understandings people have of themselves, both individually and collectively. To recognize that technologies simultaneously concern material and social production, and that both lie at the heart of social reproduction is to highlight the performative nature of technological practice (Mauss 1935; in this volume, Hoffman chapter 5, Dobres chapter 6, and Pfaffenberger chapter 7). And by taking technical performance seriously, we can begin to visualize the dynamic sociopolitical arena within which technological practice and its products materialize(d), rather than considering them separately and afterward (Ingold 1993; Reynolds 1993; Ridington 1982; see also Ingold's Foreword and Roux and Matarasso's case study in chapter 3 of this volume).

The contributors to this volume differ in the way they evaluate and explore the interplay of these multiple factors. Not only do they work at various analytic and phenomenological scales; they employ a variety of starting premises and appeal to a broad array of analytic and interpretive methodologies, from the use of culturally specific folklore and origin myths to cross-cultural economic principles derived from the study of profit margins; some focus on the macroscale context of culture contact others on the microscale of face-to-face interaction; and, they variously concern themselves with age, gender, class, and ethnicity. Rather than propose a single unified vision able to encompass these many emphases, we welcome their different attempts to infuse an explicit concern with the active, the social, and the political into what has too often been restricted to the passive, the material, and the economic.

A REVITALIZATION IN TECHNOLOGY STUDIES

In the past two decades the study of technology has been revitalized as a result of methodological advances, empirical understandings, and theoretical breakthroughs coming from several intellectual directions. In particular, experimentation and replicative studies, often grounded in ethnoarchaeological research, have revolutionized material understandings of ancient technologies. The French analytic known as *chaîne opératoire* figures prominently in many of these advances (in this volume, Roux and Matarasso chapter 3, Dobres chapter 6, and Wake chapter 9). Simultaneously, but independently, Anglo-Americans have developed and pursued the notion of behavioral chain analysis (after Schiffer 1975, 1992). These analytic and methodological advances in the study of artifact "life histories" complement a revived interest in the so-called biography of things being explored (once again) in contemporary culture studies (superb examples include Appadurai 1986; Kopytoff 1986; MacKenzie 1991; Stark 1998).

Although it is impossible to provide a detailed discussion here, current research on ancient material-processing techniques owes an equally large debt to the detailed pursuit of questions concerned with the development and organization of specialized craft production, which is especially important to understanding the origins and development of complex societies (see, for example, Costin 1991; Knapp 1990; Nelson 1991; Perlès 1989, 1992; Roux and Corbetta 1990; Sinopoli 1988; Tosi 1984; in this volume, Roux and Matarasso chapter 3, Hoffman chapter 5). Similarly, exchange studies now rely heavily and with great success on the sophisticated locational sourcing of the lithic and mineral resources used in prehistory to track the movement of materials and products through differently organized short and long-distance trade networks (e.g., Brumfiel and Earle 1987; Gale and Stos-Gale 1982; Hallam et al. 1976; Manolakakis 1994; Torrence 1986). As well, a new age of technology studies has revitalized the question of innovation. Many case-specific studies demonstrate with improved precision the historic constitution and sociopolitical underpinnings of macroscale technological change (in particular, Rice 1981; case studies in Costin et al. 1989; van der Leeuw and Torrence 1989). In these and related areas too numerous to detail, materials science studies have emerged as a major partner in archaeological research on ancient technology, thereby making significant contributions to middle-range research (for but one recent example, see Carr and Neitzel 1994).

FROM WORLD VIEWS AND TECHNOLOGICAL STYLE TO PRACTICE AND AGENCY

A major theoretical advance contributing significantly to the revitalization of technology studies in anthropology and archaeology, and the one taken up in this volume, has been fueled by research exploring the structural relationship between cultural paradigms, mental constructs, shared world views—or what Lechtman in her Afterword calls "ethnocategories"—and material technological practice. This work was pioneered by Lechtman (1977, 1984) and, more recently, Lemonnier (1986, 1989; overview in Dobres and Hoffman 1994:217–221). We see this research domain as one of the most promising in decades, and numerous case studies demonstrate unequivocally what Childe (1956:1) understood some forty years ago: that material objects and technologies are "concrete expressions and embodiments of human thought and ideas." Current research on technological world views is especially persuasive because, from its inception, it has been anchored in a significantly robust body of general theories of representation and symbolism (Lemonnier 1986). In addition, research on the ways cultural logic drives technological systems and technical choice takes advantage of methodological advances in the rigorous scientific analysis of archaeological materials, merging it with the many processual dimensions of technology and society briefly mentioned above (Lechtman 1994; Lechtman and Steinberg 1979; examples in Childs 1986; Childs and Killick 1993; Glumac 1991; Gordon and Killick 1993; Hosler 1994; Lechtman 1984). Most important as far as we are concerned, research on ancient sociotechnical world views has finally broken the silence on what was once dismissed as paleopsychology (Binford 1965:204; Leach 1973).

This book demonstrates the interrelationship between the materiality and sociality of the productive process, the sociopolitical dynamics of such endeavors, and the place of collective world views in both shaping and giving meaning to them. What researchers are only now beginning to comprehend about this recursive relationship is that "technologies are also particular sorts of cultural phenomena that reflect cultural preoccupations and express them in the very style of the technology itself" (Lechtman and Steinberg 1979:139). As numerous case studies now demonstrate in detail, "social logics unrelated to technology may weigh heavily on the evolution of technological systems" (Lemonnier 1993:2), and "at any given time, techniques form the backdrop of people's material life as well as part of their systems of meaning" (Lemonnier 1993:27). This perspective can be traced to Lechtman's (1977) pivotal concept of technological style (re-articulated and further developed in her Afterword to this volume), which has done so much to forge behavioral and material links between once separate concerns with cognition, style, and technology.

Compelling material culture studies undertaken by sociocultural anthropologists, culture studies researchers, and archaeologists demonstrate the multifaceted ways that historically grounded systems of cultural logic structure and make tangible fundamental principles underlying people's everyday activities of producing, using, and giving meaning to material objects. For archaeology in particular, these studies have succeeded in "recovering mind" (Leone 1982) through the rigorous and empirically grounded study of technical attributes and technical patterning. Equally important, they have shown that symbols, cognitive processes, and world views are central to and inseparable from people's understanding of the physical properties of materials.

Although the contributors to this volume premise their work on the dialectic of producing artifacts and reproducing cultural logic, each takes off in different directions to consider more explicitly the practice, politics, and agential dynamics instantiated in particularly structured technological styles of making, using, and living in the material world.

PUSHING AT THE EDGES

The preceding discussion notwithstanding, it is clear to us that a "world view" approach to technology and technological style does not address the essential question of *how* structural principles are expressed and reaffirmed, much less undergo change, through the interaction of technical actors with their material world. In an earlier work (Dobres and Hoffman 1994), we confronted this question directly, arguing that certain implications of this perspective remain unexplored, despite the major steps forward in recognizing the inseparability of social representations, cultural logic, and technology. In particular, we argued that the theoretical foundations for an anthropology of technology lag behind methodological advances.

This volume is premised on the position that a "world view" approach that does not take into account the everyday social experiences and political strategies of technical agents undertaking productive activities cannot satisfactorily explain *how* symbols are manifest in technique, *how* technology reaffirms fundamental social representations, or *how* such integrated sociosymbolic and material systems might change over time. In this sense, sociotechnical agency and the political dimensions of technologies are surely processual concerns. Yet the linchpin in any human technology—people (or, perhaps more properly, human agency)—has been too long absent from most archaeological considerations and far too many sociocultural studies as well.

Our 1994 essay stressed that prevailing accounts of technological systems that overlook or discount practice and agency as major contributing factors (along with world views) are untenable. Our point was not to rally a search for the identities of specific technical actors past or present, or to promote "free will" accounts for technical innovation and invention. Even *if* the empirical subject of inquiry is a single technician or the material products of an individual technical act (where such ephemeral traces preserve), the focus of attention needs to be placed squarely on the wider community of relations in which such activities took place, and on the contribution of microscale social interaction to larger sociotechnical dynamics (after Ingold 1988, 1993; Reynolds 1993). Even at the scale of the individual artifact or trace of a technical activity, products and technical agents are forever situated within social communities, systems of value, and historical conditions. No matter how tempting, we maintain that they should not be extricated from these antecedent contexts for heuristic convenience or methodological control. These larger frames of reference are an integral part of the antecedent social and material conditions giving rise to particularly structured material practices (Dobres and Hoffman 1994:222–226; prime example in Moore 1986).

As we argued in 1994, the lack of sustained attention to the social agency of technological practice, past or present, is due in large part to two unnecessary conceptual roadblocks. The first of these is epistemological, the second theoretical in nature. With the specter of Hawkes's (1954) epistemological hierarchy still dominating the discipline, archaeologists (in particular) have been more intent on producing trustworthy middle-range inferences about the supposedly more knowable aspects of ancient human behaviors than on investigating the intangible dynamics of social agency that lie at the heart of cultural processes. In the case of the second roadblock, we hold that a major reason why the study of technological agency has not received as much attention as the symbolic dimensions of technique is in large part because there has been no sustained attempt to apply a robust theory of social agency to the question concerning technology (Dobres and Hoffman 1994:222–226; with apologies to Heidegger 1977; but see Dietler and Herbich 1998).

The profound silence on the topic of agency in material cultural studies goes far beyond the field of technology, and even archaeology. For some time now, social scientists pursuing many different research agendas have described their increasing frustration with the conceptual and methodological "black box" effectively sealing off from sustained consideration the anthropological dynamic we call agency (see Archer 1988, 1995; Bourdieu 1977; Brumfiel 1991; Cowgill 1975;

Giddens 1979, 1984; Hodder 1986; Shanks and Tilley 1987; Sztompka 1991, 1994; Tringham 1991). That silence is especially troubling where technology studies are concerned: we remain unconvinced that there can ever be an adequate anthropological or processual understanding of technology, technological systems, or technological change unless technical agents are a central and essential part of our explanatory and interpretive models.

To promote more awareness and concern with sociotechnical agency and technological practice, our 1994 essay focused on four analytic factors: scale, context, materiality, and social theory. First, technology studies require more explicit concern with technological practice at the *microscale.* By microscale, we mean the phenomenological scale at which the experience of everyday artifact manufacture, use, and repair by individuals and small collectivities becomes meaningful social practice. At this scale are fundamental social, material, and antecedent *contexts* within which technologies take physical shape and acquire their social value. Such interpersonal social arenas are an important complement to macroscale perspectives, which focus on less intimate and immediately perceived contexts, such as ecological circumstances and evolutionary processes. *Materiality,* too, plays an important role in technological practice. It is therefore essential that researchers attend to the physical properties of particular technologies while bringing to prominence the microscale dynamics of agency and meaning allowing their manipulation and modification.

What is needed above all, however, is a body of *social theory* that can relate technical knowledge and material action to social knowledge and political action. Contemporary theories of social agency are especially attractive in this regard because they demonstrate how intimate and everyday routines constitute the "stuff" of larger cultural processes (especially, Archer 1988, 1995; Bourdieu 1977; de Certeau 1984; Giddens 1979, 1984; Sztompka 1991, 1994). For all its focus on routines and habits, so-called practice, or agency, theory is actor-centered, and allows social agents at least partial awareness of what they do. Although individuals may not always be successful, theories of practice give at least a "mediated understanding" of how to act under particular circumstances and can do much to historicize and humanize our understanding of technology. At the same time, theories of agency are practically mute on the active role of material culture and technological endeavors in everyday expressions of self- and group-interest.

Across the board, students of material culture are beginning to discover how politics and power are particularly salient expressions of social agency enacted in and through material and technical means (examples include Cockburn 1992; Conkey 1991; Dobres 1995b; Herbert 1993; Ingold 1993; Law 1991; Lechtman 1993; Winner 1986b). As several chapters in this volume show, technologies, technical acts, and technical knowledge express, contest, and mediate interpersonal and group interests as much as they reaffirm normative world views. These case studies of technopolitics are firmly embedded in social theory but push beyond it by attending to material matters. In particular, the chapters by Childs (2), Hoffman (5), Dobres (6), Pfaffenberger (7), and Wake (9) demonstrate how context specific and materially grounded expressions of self- and group-interest, along with the enactment of interpersonal relations, are necessarily implicated in technical performance at multiple phenomenological scales, from the individual to the collective. Ridington's chapter (8), in particular, explores how social power turns on the construction and differential control of technical knowledge and shows how technique (in both its tangible and intangible dimensions) can serve as an empowering system of knowledge helping actors situate themselves in their sociomaterial world.

In our view, one of the central problems with contemporary theories of material culture and social agency is that they overlook the nuanced ways world views come to be represented in and by the specific material choices technicians make. Furthermore, they often fail to consider how cultural logic structures sociotechnical activities in particular times and places, and how mental schemas are manifest in technological end-products and in the way they are used by agents with a variety of personal and collective agendas. The contributors to this volume grapple with these concerns, albeit in different ways and with different conclusions. Lechtman's Afterword exemplifies how much this general approach can accomplish when grounded in detailed materials study. For example, the connections Childs finds in chapter 2 between Toro technical practice and normative gender cosmologies, and the relationship Larick detects in chapter 4 in changing gender configurations and the built environment of colonial New England, are not congruent—nor should they be if the agency of technical systems is, in part, historically and culturally contingent. Similarly, Roux and Matarasso (chapter 3), Pfaffenberger (chapter 7), and Wake (chapter 9) show that normative attitudes about the "right" and "wrong" ways to make and use material culture (be it stone beads, storage rooms and canoes, or fishing implements) have underlying processes in common that, while responsive to the particular media employed, are sensitive to the culturally defined goals of the technicians involved and their culturally specific cosmologies. Finally, Hoffman (chapter 5), Dobres (chapter 6), and Ridington (chapter 8) suggest that the interplay of technical acts, products, and knowledge can work to promote the self-interested agendas of technical agents in the face of larger (that is, constraining) social and material structures in which all technical systems are situated.

Given the preliminary and exploratory nature of these studies, we see no value in establishing an overarching cross-cultural definition of what "should" constitute technopolitics and technical agency for all times, places, contexts, and media. Not only have we barely begun to identify the relevant issues involved; we remain unconvinced that some umbrella-like theory could ever be rigorous enough to explicate the context-specific dynamics lying at the heart of technological practice, politics, and world views (Dobres and Hoffman 1994:247–248). That caveat aside, some key facets of technological practice and agency appear throughout this volume: technical knowledge; manual skill and *savoir-faire;* social relations of production; the intersection of social identity with the "life history" of artifacts; value systems associated with technical practice, knowledge, and end-products; and political expressions of self- and group-interest instantiated through material means. The materiality of technology is necessarily implicated in all of this; hence research must be especially careful not to turn away from explicit concern with the physical dimensions of resources, technique, and end-product. The efficacy of technology necessarily lies *within* the dialectic of the material, social, political, and symbolic.

PHILOSOPHICAL AND HISTORICAL REFLECTIONS ON THE SILENCE OF TECHNICAL AGENCY

As philosophers, historians, and sociologists of technology have long noted, the *conceptual* distance that we imagine between ourselves and our Western industrial technologies, which Marx called alienation and fetishism, is little more than our own world view, ideologically congruent with capitalism, modernity, and industrialism (Habermas 1970; Heidegger 1977; Latour 1993; Marcuse 1968; Mitcham 1994; Pacey 1983; Winner 1977). Unwittingly, archaeologists have employed such presentist and mechanistic visions in their research programs (Dobres 1995a, forthcoming; Hoffman 1993; Pfaffenberger 1988, 1992). Implicitly, they have retrofitted into prehistory the same alienated relationship we are said to have with modern industrial technologies, a relationship captured in the notion of technological determinism. Among other things, technological determinism supports the belief that once created, technologies take on a life of their own and exist outside the body politic (Haraway 1991; Latour 1993; Winner 1977). As long as we continue to project uncritically modernist and instrumentalist notions of technology onto the technological practices and world views of other people (past or present), we will continue to bracket off as a second-order dynamic the very dialectic that makes technologies possible, namely, that technology and society are *inseparable.* If we continue to approach the study of ancient technologies the way we live our own lives, steeped in deterministic attitudes about technology divorced from the social body, we will forever separate what are essentially inseparable facets of a single, albeit highly complex, phenomenon (Dobres 1995a, 1995c; Mumford 1967). This volume offers a partial corrective to this situation.

BEYOND DISCIPLINARY BOUNDARIES

Why bring together in one volume essays by sociocultural anthropologists, ethnoarchaeologists, and archaeologists? The purpose of this volume is not to present different aspects of the technology-society "equation," but to stress the value of understanding technology from complementary perspectives, in the hope of arriving at a richer and more nuanced understanding of the whole (Miller et al. 1996). Although sociocultural anthropologists and ethnoarchaeologists have the advantage of directly observing ongoing technological systems (where intangible social dynamics are said to be more accessible and understandable), it is archaeologists who have developed the most sophisticated and creative methodologies to study the empirical and physical dimensions of material culture and the productive process. We need each other to create meaningful links between statics and dynamics, between tangibles and intangibles, between artifacts and artifice, and between products, processes, and people.

While our goal in this volume is to show that the social, symbolic, political, and material aspects of technological practice are indeed inseparable, the discussion is most easily followed if divided into subsections: namely, technological practice, politics, and technological world views. This strategy, however, should not be confused with our embracing a concept of culture partitioned into separate "spheres" such as social organization, economics, politics, technology, and belief systems, connected by the thinnest and straightest of lines. For in the very act of separating then reconnecting these dots, archaeologists and sociocultural anthropologists often lose sight of people and the seemingly intangible dynamics that weave these processes together (Brumfiel 1991; Dobres forthcoming; Latour 1993).

Clearly, it is time to reconceptualize technology as an inseparable whole that cannot and should not be divided into constituent parts. At best, it can be turned and rotated on its axis to provide different views of its totality. With that in mind, we have purposefully juxtaposed material technologies, times, places, historical circumstances, and methodological emphases in each section. Our intent is not to create a methodological road map or to suggest some unifying interpretive or explanatory perspective "best" suited to understanding technological dynamics through time and space. Rather, we want to give a prominent place to the *nexus* of factors involved, thereby providing a forum in which these issues can be explored by author and reader, alike.

CONCLUSIONS AND BEGINNINGS

One of our primary objectives in this volume is to add a new dimension to the study of technological world views and style: an explicit and informed concern with practice and agency. To accomplish this, we also try to put a human face on technology more generally. At the same time, we want to make clear that not all contributors agree with, or necessarily accept, all the positions discussed in this chapter. The subject tackled in this volume—technology—is simply too complex to expect consensus as to a single key dynamic or even a methodology for studying it. Because of the difficulty in simultaneously studying the sociality and materiality of technological practice, we are compelled to explore as many promising avenues as possible. We suspect that multifaceted attempts—especially those employing multiple lines of evidence—will have the best chance of success.

Although we promote a more historical and particularistic approach to the study of technology, we are not suggesting that researchers should abandon cross-cultural studies or that they should stop modeling evolutionary-scale processes of wholesale technological change. In chapter 3, Roux and Matarasso provide an important example of how empirically grounded research conducted at microanalytic scales can shed light on macroscale questions, in their case concerning the origin of specialized craft production and state formation. What we do want to promote, show the need for, and value of is *tolerance* for alternative scales of analysis and research agendas. Because technology is a multifaceted cultural practice, it can and should be studied from a variety of perspectives, especially those bridging processual and postprocessual paradigms (Marquardt 1992; Preucel 1991). From this, a more coherent, albeit extremely complex, anthropological understanding of technology may one day emerge. But until our frameworks move beyond the orthodox privileging of practical over cultural reason (Sahlins 1976) and include explicit concern with sociopolitical, agential, and symbolic aspects of such practices, we will forever have top-down models that see Nature, Natural Selection, or some other extrasocial dynamic dictating and structuring what is (and was) a decidedly cultural phenomenon (Ingold 1995). Though linear analyses and their resulting explanatory models may be heuristically satisfying, they offer only a partial understanding of an exceedingly complex cultural phenomenon that requires far more sophisticated study.

REFERENCES CITED

Appadurai, A. (editor)

1986 *The Social Life of Things.* Cambridge University Press, New York.

Archer, M. S.

1988 *Culture and Agency: The Place of Culture in Social Theory.* Cambridge University Press, Cambridge.

1995 *Realist Social Theory: The Morphogenetic Approach.* Cambridge University Press, Cambridge.

Binford, L. R.

1965 Archaeological Systematics and the Study of Culture Process. *American Antiquity* 31(2):203–210.

Bourdieu, P.

1977 *Outline of a Theory of Practice.* Translated by R. Nice. Cambridge University Press, Cambridge.

Brumfiel, E. M.

1991 Distinguished Lecture in Archaeology: Breaking and Entering the Ecosystem: Gender, Class, and Faction Steal the Show. *American Anthropologist* 94(3):551–567.

Brumfiel, E. M., and T. K. Earle (editors)

1987 *Specialization, Exchange and Complex Societies.* Cambridge University Press, Cambridge.

Carr, C., and J. E. Neitzel (editors)

1994 *Style, Society and Person: Archaeological and Ethnological Perspectives.* Plenum, New York.

Childe, V. G.

1956 *Society and Knowledge.* Harper and Brothers, New York.

Childs, S. T.

1986 *Style in Technology: A View of African Early Iron Age Iron Smelting through Its Refractory Ceramics.* Unpublished Ph.D. dissertation, Department of Anthropology, Boston University.

Childs, S. T., and D. Killick

1993 Indigenous African Metallurgy: Nature and Culture. *Annual Review of Anthropology* 22:317–337.

Cockburn, C.

1992 Technology, Production, and Power. In *Inventing Women: Science, Technology, and Gender,* edited by G. Kirkup and L. Smith Keller, pp. 196–211. Polity Press, Cambridge.

Conkey, M. W.

1991 Contexts of Action, Contexts for Power: Material Culture and Gender in the Magdalenian. In *Engendering Archaeology: Women and Prehistory,* edited by J. M. Gero and M. W. Conkey, pp. 57–92. Blackwell, Oxford.

Costin, C. L.

1991 Craft Specialization: Issues in Defining, Documenting, and Explaining the Organization of Production. *Archaeological Method and Theory* 3:1–56.

Costin, C. L., T. K. Earle, B. Owen, and G. Russell

1989 The Impact of Inca Conquest on Local Technology in the Upper Mantaro Valley, Peru. In *"What's New?" A Closer Look at the Process of Innovation,* edited by S. E. van der Leeuw and R. Torrence, pp. 107–139. Unwin Hyman, London.

Cowgill, G. L.

1975 On Causes and Consequences of Ancient and Modern Population Changes. *American Anthropologist* 77(3):505–525.

de Certeau, M.

1984 *Practice of Everyday Life*. University of California Press, Berkeley.

Dietler, M., and I. Herbich

1998 *Habitus,* Techniques, Style: An Integrated Approach to the Social Understanding of Material Culture and Boundaries. In *The Archaeology of Social Boundaries,* edited by M. T. Stark, pp. 232–263. Smithsonian Institution Press, Washington, D.C.

Dobres, M-A.

1995a *Gender in the Making: Late Magdalenian Social Relations of Production in the French Midi-Pyrénées.* Ph.D. dissertation, Department of Anthropology, University of California at Berkeley. University Microfilms, Ann Arbor.

1995b Gender and Prehistoric Technology: On the Social Agency of Technical Strategies. *World Archaeology* 27(1):25–49.

1995c Prehistoric Cyborgs!? Or, How to Weave Ancient Technosocial Webs in the Present. Paper presented at the Annual Meetings of the American Anthropological Association, Washington, D.C.

Forthcoming *Technology and Social Agency: Outlining an Anthropological Framework for Archaeology.* Blackwell, Oxford.

Dobres, M-A., and C. R. Hoffman

1994 Social Agency and the Dynamics of Prehistoric Technology. *Journal of Archaeological Method and Theory* 1(3):211–258.

Gale, N. H., and Z. Stos-Gale

1982 Bronze Age Copper Sources in the Mediterranean: A New Approach. *Science* 216:11–19.

Giddens, A.

1979 *Central Problems in Social Theory: Action, Structure, and Contradiction in Social Analysis.* University of California Press, Berkeley.

1984 *The Constitution of Society: Outline of a Theory of Structuration.* University of California Press, Berkeley.

Glumac, P. D. (editor)

1991 *Recent Trends in Archaeometallurgical Research.* MASCA Research Papers in Science and Archaeology 8(1). MASCA, The University of Pennsylvania, Philadelphia.

Gordon, R. B., and D. Killick

1993 Adaptation of Technology to Culture and Environment: Bloomery Iron Smelting in America and Africa. *Technique and Culture* 34(2):243–270.

Habermas, J.

1970 Technology and Science as "Ideology." In *Toward a Rational Society: Student Protest, Science, and Politics,* pp. 81–122. Beacon Press, Boston.

Hallam, B., S. Warren, and C. Renfrew

1976 West Mediterranean Obsidian. *Proceedings of the Prehistoric Society* 42:85–110.

Haraway, D.

1991 A Cyborg Manifesto: Science, Technology, and Socialist-Feminism in the Late

Twentieth Century. In *Simians, Cyborgs, and Women: The Reinvention of Nature,* pp. 149–181. Routledge, New York.

Hawkes, C. F.
1954 Archaeological Theory and Method: Some Suggestions from the Old World. *American Anthropologist.* 56:155–168.

Heidegger, M.
1977 *The Question Concerning Technology.* Translated by W. Lovitt. Garland Publishers, New York.

Herbert, E. W.
1993 *Iron, Gender, and Power: Rituals of Transformation in African Societies.* Indiana University Press, Bloomington.

Hodder, I.
1986 *Reading The Past.* Cambridge University Press, Cambridge.

Hoffman, C. R.
1993 *The Social and Technological Dimensions of Copper Age and Bronze Age Metallurgy in Mallorca, Spain.* Unpublished Ph.D. dissertation, Department of Anthropology, University of California at Berkeley.

Hosler, D.
1994 *The Sounds and Colors of Power: The Sacred Metallurgical Technology of Ancient West Mexico.* MIT Press, Cambridge, Mass.

Hughes, T. P.
1979 The Electrification of America: The System Builders. *Technology and Culture* 20(1):124–162.

Ingold, T.
1988 Tools, Minds, and Machines: An Excursion in the Philosophy of Technology. *Techniques et Culture* 12:151–176.
1993 Tool-Use, Sociality and Intelligence. In *Tools, Language, and Cognition in Human Evolution,* edited by K. R. Gibson and T. Ingold, pp. 429–445. Cambridge University Press, Cambridge.
1995 "People Like Us": The Concept of the Anatomically Modern Human. *Cultural Dynamics* 7(2):187–214.

Knapp, A. B.
1990 Production, Location, and Integration in Bronze Age Cyprus. *Current Anthropology* 31(2):147–176.

Kopytoff, I.
1986 The Cultural Biography of Things: Commodization as a Process. In *The Social Life of Things,* edited by A. Appadurai, pp. 64–91. Cambridge University Press, New York.

Latour, B.
1993 *We Have Never Been Modern.* Harvard University Press, Cambridge, Mass.

Law, J. (editor)

1991 *A Sociology of Monsters: Essays on Power, Technology, and Domination.* Routledge, New York.

Leach, E. R.

1973 Concluding Address. In *Explanation of Culture Change: Models in Prehistory,* edited by C. Renfrew, pp. 761–771. Duckworth, London.

Lechtman, H.

1977 Style in Technology: Some Early Thoughts. In *Material Culture: Styles, Organization, and Dynamics of Technology,* edited by H. Lechtman and R. S. Merrill, pp. 3–20. West Publishing Co., St. Paul, Minn.

1984 Andean Value Systems and the Development of Prehistoric Metallurgy. *Technology and Culture* 15(1):1–36.

1993 Technologies of Power: The Andean Case. In *Configurations of Power in Complex Societies,* edited by J. S. Henderson and P. J. Netherly, pp. 244–280. Cornell University Press, Ithaca, N.Y.

1994 Materials Sciences of Material Culture: Examples from the Andean Past. In *Archaeometry of Pre-Columbian Sites and Artifacts,* edited by D. A. Scott and P. Meyers, pp. 3–11. Getty Conservation Institute, Los Angeles.

Lechtman, H., and A. Steinberg

1979 The History of Technology: An Anthropological Perspective. In *History and Philosophy of Technology,* edited by G. Bugliarello and D. B. Doner, pp. 135–160. University of Illinois Press, Urbana.

Lemonnier, P.

1986 The Study of Material Culture Today: Towards an Anthropology of Technical Systems. *Journal of Anthropological Archaeology* 5:147–186.

1989 Bark Capes, Arrowheads, and Concords: On Social Representations of Technology. In *The Meaning of Things: Material Culture and Symbolic Expression,* edited by I. Hodder, pp. 156–171. Unwin Hyman, London.

1993 Introduction. In *Technological Choices: Transformation in Material Cultures Since the Neolithic,* edited by P. Lemonnier, pp. 1–35. Routledge, London.

Leone, M. P.

1982 Some Opinions about Recovering Mind. *American Anthropologist* 47:742–760.

MacKenzie, M. A.

1991 *Androgynous Objects: String Bags and Gender in Central New Guinea.* Harwood Academic Publishers, Chur, Switzerland.

Manolakakis, L.

1994 *La production des outils en silex dans les sociétés hiérarchisées de l'enéolithique en Bulgaire: Evolution, traditions culturelle et organisation sociales.* Doctorat, Université de Paris I, Panthéon Sorbonne.

Marcuse, H.

1968 Industrialization and Capitalism in the Work of Max Weber. In *Negations: Essays in Critical Theory,* translated by J. J. Shapiro, pp. 201–226. Beacon Press, Boston.

Marquardt, W. H.
1992 Dialectical Archaeology. *Archaeological Method and Theory* 4:101–140.

Mauss, M.
1935 Les techniques du corps. In *Sociologie et Psychologie, Parts II–VI.* (Reprinted in: *Sociologie et Anthropologie,* pp. 365–386. Presses Universitaires de France, Paris, 1950. Also in *Sociology and Psychology: Essays of Marcel Mauss,* translated by B. Brewster, pp. 97–123. Routledge and Kegan Paul, London, 1979.)

Miller, D., and C. Y. Tilley
1996 Editorial. *Journal of Material Culture* 1(1):5–14.

Mitcham, C.
1994 *Thinking Through Technology: The Path between Engineering and Philosophy.* University of Chicago Press, Chicago.

Moore, H. L.
1986 *Space, Text, and Gender: An Anthropological Study of the Marakwet of Kenya.* Cambridge University Press, Cambridge.

Mumford, L.
1967 *Technics and Human Development: The Myth of the Machine.* Vol. 1. Harcourt Brace Jovanovich, New York.

Nelson, M. C.
1991 The Study of Technological Organization. *Archaeological Method and Theory* 3:57–100.

Pacey, A.
1983 *The Culture of Technology.* MIT Press, Cambridge, Mass.

Perlès, C.
1989 *From Stone Procurement to Neolithic Society.* David Skomp Distinguished Lecture in Anthropology, Indiana University, Bloomington.
1992 Organization of Production and Systems of Exchange in the Neolithic of Greece. *Journal of Mediterranean Archaeology* 5(2):115–164.

Pfaffenberger, B.
1988 Fetishized Objects and Humanized Nature: Towards an Anthropology of Technology. *Man* 23:236–252.
1992 Social Anthropology of Technology. *Annual Review of Anthropology* 21:491–516.

Preucel, R. W. (editor)
1991 *Processual and Postprocessual Archaeologies: Multiple Ways of Knowing the Past.* Occasional Paper No. 10. Center for Archaeological Investigations, Southern Illinois University, Carbondale.

Pursell, C. W., Jr.
1985 The History of Technology and the Study of Material Culture. In *Material Culture: A Research Guide,* edited by T. J. Schlereth, pp. 113–126. University Press of Kansas, Lawrence.

Reynolds, P. C.

1993 The Complementation Theory of Language and Tool Use. In *Tools, Language, and Cognition in Human Evolution,* edited by K. R. Gibson and T. Ingold, pp. 407–428. Cambridge University Press, Cambridge.

Rice, P. M.

1981 The Evolution of Specialized Pottery Production. *Current Anthropology* 22(3): 219–240.

Ridington, R.

1982 Technology, World View, and Adaptive Strategy in a Northern Hunting Society. *Canadian Review of Society and Anthropology* 19(4):469–481.

Roux, V., and D. Corbetta

1990 Wheel-Throwing Technique and Craft Specialization. In *The Potter's Wheel: Craft Specialization and Technical Competence,* edited by V. Roux (in collaboration with D. Corbetta), pp. 3–92. Oxford and IBH, New Delhi.

Sahlins, M. D.

1976 *Culture and Practical Reason.* University of Chicago Press, Chicago.

Schiffer, M. B.

1975 Behavioral Chain Analysis: Activities, Organization, and the Use of Space. *Fieldiana* 65:103–174.

1992 *Technological Perspectives on Behavioral Change.* University of Arizona Press, Tucson.

Shanks, M., and C. Tilley

1987 *Social Theory and Archaeology.* University of New Mexico Press, Albuquerque.

Sinopoli, C. M.

1988 The Organization of Craft Production at Vijayanagra, South India. *American Anthropologist* 90(3):580–597.

Stark, M. T. (editor)

1998 *The Archaeology of Social Boundaries.* Smithsonian Institution Press, Washington, D.C.

Sztompka, P.

1991 *Society in Action.* University of Chicago Press, Chicago.

Sztompka, P. (editor)

1994 *Agency and Structure: Reorienting Social Theory.* Gordon and Breach, Switzerland.

Torrence, R.

1986 *Production and Exchange of Stone Tools: Prehistoric Obsidian in the Aegean.* Cambridge University Press, New York.

Tosi, M.

1984 The Notion of Craft Specialization and Its Representation in the Archaeological Record of Early States in the Turanian Basin. In *Marxist Perspectives in Archaeology,* edited by M. Spriggs, pp. 22–52. Cambridge University Press, Cambridge.

Tringham, R. E.

1991 Households with Faces: The Challenge of Gender in Prehistoric Architectural Remains. In *Engendering Archaeology: Women and Prehistory,* edited by J. M. Gero and M. W. Conkey, pp. 93–131. Blackwell, Oxford.

van der Leeuw, S. E., and R. Torrence (editors)

1989 *What's New? A Closer Look at the Process of Innovation.* Unwin Hyman, London.

Winner, L. C.

1977 *Autonomous Technology: Technics-Out-of-Control as a Theme in Political Thought.* MIT Press, Cambridge, Mass.

1986a *The Whale and the Reactor: A Search for the Limits in an Age of High Technology.* University of Chicago Press, Chicago.

1986b Do Artifacts Have Politics? In *The Whale and the Reactor: A Search for the Limits in an Age of High Technology,* pp. 19–39. University of Chicago Press, Chicago.

I TECHNOLOGICAL PRACTICE

2. "After All, a Hoe Bought a Wife": The Social Dimensions of Ironworking among the Toro of East Africa

S. TERRY CHILDS

"AFTER ALL, A HOE BOUGHT A WIFE." Mzee Ndunga,[1] an elder Toro ironsmith and my chief informant, repeated several versions of this simple statement during our discussions about precolonial and early colonial ironworking in the Toro kingdom of present-day Uganda (figure 2.1). For him, the words seemed to encapsulate how and why ironworking and its actors were significant and had social stature within his society. For me, the words initially spoke about the complex nature of the dynamic called technology, the one that has been characterized as "simultaneously material, social and symbolic" (Pfaffenberger 1988:236). It became increasingly clear during our conversations in 1994 that the social actions and relationships involved in modern Toro ironworking were as important and fundamental to its existence as its materiality. The profoundly social nature of Toro ironworking is the focus of this chapter.

Ndunga acknowledged the critical importance of the material aspects of iron making through his reference to the end-product of the arduous process, the hoe. In fact, the traditional production of a hoe was the epitome of the ironworkers' careful and skillful control over various raw materials, tools, and technical knowledge. Ndunga affirmed the significance of the production process and its technical and material components when he referenced an iron-making tool in other remarks. For instance, he noted that "when a woman (wife) doesn't know the *tuyere,* she doesn't know what bought her."[2] The *tuyere,* or blowpipe, was key to the technical success of both iron smelting and forging because it forced air into the blast zones of the furnace and the forge to achieve the high temperatures necessary to work the metal. Yet it was precisely because of their relationships to

Figure 2.1. Map of Uganda showing the boundaries of the kingdoms of Toro and Bunyoro in colonial times. Adapted from Beattie (1971).

technical success that Ndunga also used the hoe and the *tuyere* as material expressions of the resulting wealth that allowed ironworkers to marry.

For archaeologists, the individual raw materials (such as charcoal, iron ore, clay), tools (such as furnace, bellows, hammers), and end-products (especially iron objects, slag) are the most tangible and lasting constituents of iron making. They provide the concrete, observable evidence of the technical processes that communities practiced across the landscape and that they might have changed over time (Childs 1996; Miller and van der Merwe 1994; Schmidt and Childs 1985; Tylecote 1986). The technical impacts on and relationships to social lives and social contexts rarely yield tangible evidence in the archaeological record. Nor do the symbolic and ideological nuances of technologies. Fortunately, a few archaeologists have recently become interested in explicitly exploring these other aspects of ancient technologies (Childs 1986, 1991, 1994; Conkey 1990; Dobres 1995; Dobres and Hoffman 1994; Hodder 1986; Ingold 1991; Lechtman 1977; Schiffer 1992; Schmidt 1983, 1996, 1997). Many often acknowledge the role of ethnoarchaeology and ethnography in helping to understand the social and ideological complexities involved.

Ndunga's comment that "a hoe bought a wife" went far beyond materiality because it spoke to the very broad social context in which ironworking and its participants play a part and to which they are intimately related. Buying and selling a hoe involves processes of evaluation, negotiation, exchange, and compensation between the smith and his customer(s) now and in the past. When a hoe was—and still is—available to buy at a market or at the workplace, the related activities and people involved in forging the hoe, as well as farming and clearing land, were clearly implied and considered. A poorly made Toro hoe was duly noted by potential customers and never bought, so the smith and his assistants had to reforge it before it was acceptable.

The influence of a culture's world view and associated value systems on decision making within a technological context (Childs and Killick 1993; David et al. 1988; Gosselain 1992; Herbert 1993; Lechtman 1977) was also underscored in Ndunga's remark. The manufacture of a hoe required more iron and more labor from a Toro smith than any other iron object. It was a valuable commodity for Toro families whose survival depended on hoe-based agriculture and for the women who tended the gardens. It was also valuable to the smith who, after making it, could exchange it for as much as a healthy young cow. Or, he could decide to use the hoe himself as bride price for a wife, a most valued individual for a Toro man.[3] Without a wife, a man could not—and still cannot—take the first steps on his journey toward becoming a full and valued member of society. It was only when his wife provided him with children that he gained full social status and created a legacy. The more wives and children a man had, the greater his prestige and status within the community (Taylor 1962). Thus the value of a hoe, as well as the technical skill required to smelt the iron ore and forge it properly, were more than practical means to an end. They were also laden with cultural significance that helped structure the material act of hoe production.

As Herbert (1993) has shown so well for African ironworking in general, age and gender framed many of the web of decisions made during technical and related activities. Similarly, older Toro men controlled the ironworking process by using culturally sanctioned rules to limit participation and restrict access to its secrets, ultimately for their own benefit. These rules were founded on the Toro world view of human relations between young and old, men and women, and the living and dead. In precolonial days, as today, Toro society was patriarchal, and women had fewer rights than men; the longer people lived, the more social stature they accrued; spirits and ancestors had considerable influence over the living as a result of the experiences and transformations they had undergone (Beattie 1971; Taylor 1962). Ironworkers incorporated rituals and material symbols, including certain ironworking tools, into their technical repertoire of activities to enact, express, and more subtly reinforce these values and principles. Even the organization of their labor pool was structured by this complex of factors.

Ndunga's ultimate message was that those who controlled the iron-making technology had greater access to many of the things that were highly valued in Toro society, particularly those related to productivity, social status, and prosperity. A significant factor in an ironworker's success was the wide network of social relations he both took advantage of and influenced.

This, then, is the context for understanding in detail the interplay of social and technical factors involved in Toro iron production. This interplay structured and gave meaning to its practice through the manipulation and use of culturally accepted values, principles, and rules of behavior. As discussed next, numerous, primarily technical, roles were negotiated and filled during the three stages of Toro iron making: mining and preparing the ore, smelting the ore into raw metal, and forging the raw metal into objects, along with the broader social dynamics played out during each technical stage, including ritual, symbolic, and social gatherings of non-ironworkers. Membership in a family, clan, village, interclan cadre of ironworkers, and the Toro polity determined who was included and excluded from each phase of ironworking and its related activities. Despite the normative rules and proscriptions involved, there are occasions for social opportunism by non-ironworkers in these technological contexts, as this chapter shows. I conclude this consideration with some general thoughts about how ethnographic and ethnoarchaeological perspectives can bring considerable insight to studies of past technologies from an explicitly social perspective.

THE ACTORS AND CONTEXTS OF TORO IRONWORKING

It is ironic, and perhaps revealing about scholarly priorities and interests, that the bowl type of iron-smelting furnace used by the Toro has been labeled the "simplest," "crudest," or "least sophisticated" of three or four primary furnace types found in Africa (Cline 1937; Kense 1983; van der Merwe 1980). Scholars have focused on the small size of the bowl furnace, its inability to tap the resulting slag, its small number of bellows (one or two), and therefore its relatively small yield of smelted iron. Rarely discussed is the potentially high quality of the iron product or the complex network of people that might be involved. Many also tacitly assume that if a society uses a "simple" smelting furnace, the entire ironworking process is, by extension, "simple." In the Toro case, the number of roles and people involved in the overall iron-making process was far from simple. In precolonial and early colonial times, the principal jobs in the suite of operations required were filled mostly by men.

There were three primary roles in mining. The most important was called the *omujumbuzi,* the person who discovered a new hill of ore and who became its instant owner.[4] He was the fortunate one in a search party of five to ten ironworkers who all sought to find ore first. Once an ore source was identified, the discoverer negotiated with two groups of workers to mine the ore while he supervised. The *abahaige,* usually two to four men who spelled one another, dug the raw ore from the deep narrow mine shafts in total darkness. Usually, two *fundi* sat at the top of the pits, hauled up basketloads of material, and identified and separated the ore from the mud and other stones. These workers then brought the mined ore to the compound of the discoverer and owner of the ore, where it was carefully dried under his direction.

When the ore was properly dried and ready for smelting in a furnace, only two principal people were involved: the head smelter, loosely called the *omuhesi,* or smith, by Ndunga and his colleagues, and the *omujugusi* who pumped air into the furnace through a combination of a pair of bellows and a *tuyere* (blowpipe). The services of a *rubumbi* (potter) were sometimes required to make the important fire-hardened clay bellows pots and *tuyere.* Although many Toro women made clay cooking pots, the *rubumbi* of bellows and *tuyeres* were always men. An *omutezi ba bamakara* might be asked to make charcoal if, for some reason, the men conducting the smelt were unable to do so. Finally, one of the smelters' wives often brought them food during the many hours of arduous work.

At the end of the smelt, two to four *abahereza,* or servers, pulled the bloom, the mass of raw iron mixed with waste slag, out of the furnace with long sticks.[5] The *omwasi w'ebyoma,* or splitter, then used an axe to cut the bloom into pieces while it was still soft from the heat of the furnace.

The *omuhesi mukuru* was the head smith at the forge where objects were made out of bloom or repaired, or where worn-out items were reforged into other types of objects. Today, Toro smiths primarily use truck and Land Rover springs for their raw material instead of bloom. Two other persons at the forge were the *omurubiki,* who pounded the raw iron into flat pieces, and the *omujugusi,* the bellower. The smith's wife often supplied water and food for the workers in this context.

Another role, vital to all three stages of iron making, did not require or use any technical knowledge. It was that of the *nyakatagara,* or spirit medium, who communicated with the family or lineage *omuchwezi* (deity) and performed a number of rituals related to health, childbirth, and various life choices. In the context of ironworking, rituals were performed to influence the successful outcome of all the technical activities. Whereas most of the roles described above were filled by men, the Toro *nyakatagara* was usually a woman (Beattie 1968a; Taylor

1962). Women seem to have been preferred for this role because of their powers related to fertility, procreation, and abundance.[6] Associations with creation and productivity were, in fact, important to the entire ironworking process.

Two points should be emphasized here. First, although all of these roles required a certain degree of expertise and knowledge, some demanded more than others. Second, none of the roles were fixed by a set of static, proscriptive rules or expectations. The same person could fill more than one role, perhaps a miner on one day and a bellower during an iron smelt on another. Many roles were also negotiated and contested, both in terms of who filled them and the amount of compensation demanded and paid for the work. The master smith (*omuhesi omukuru*), for example, was usually the most experienced and senior member of a family of smiths. When he chose to retire or prior to his death, however, the master smith might name someone other than the obvious choice (his eldest son) as his successor. A bypassed brother, nephew, or cousin of the successor might then, in reaction to the designation, move elsewhere to establish his own forge. On the other hand, the person who discovered an ore source was downright lucky; all members of the search group had similar expertise in finding ore, but only one could be the first to find the stones and reap the subsequent benefits.

THE SOCIAL, IDEOLOGICAL, AND MATERIAL DYNAMICS OF TORO IRONWORKING

The technical processes of ironworking—the physical identification of good ore, the design of the smelting furnace and its proper preparation, the expertise in putting together the bellows and pumping them; or the completion of a complicated weld during the shaping of an iron object—demanded considerable knowledge and skills. These were acquired by Toro ironworkers over the course of long apprenticeships. As mentioned earlier, however, equally important to the production process were social networks of people, which provided support and motivation; social values and principles, which guided decision making and rules of behavior; and nontechnical knowledge, which helped ensure success. More can be said of the three stages of Toro ironworking in order to highlight the critical interrelationships between these technical and nontechnical factors.

Mining

In general, the details of traditional African mining are poorly known and are often presented in the literature as a relatively mundane and simple pursuit of a critical raw material by knowledgeable male ironworkers (Cline 1937). When a high-volume smelting industry developed, however, miners sometimes had to recruit help from others, including female relatives (Herbert 1998). In other cases, the miners interacted with a spirit who was ritually thanked for guarding the ore or was ritually appeased when the ore was mined and taken away (Herbert 1993). The network of social dynamics between miners, nonminers, and nonliving spirits are evident in such examples, but little attempt has been made to understand the implications of their interactions.

By contrast, the simple pursuit of ore during Toro mining activities was far from mundane, as I learned during my fieldwork. The discovery of a new mine and mining itself initiated a complex web of social relationships and rules of social behavior that were played out during the subsequent stages of iron smelting and smithing, as well as during daily life.

The search for ore was carried out exclusively by Toro men, primarily by married, older men. To find a new hill of iron ore meant wealth and prestige within one's family, clan, and village. It was therefore pursued very seriously. Men searching for ore formed groups of six or more, largely because sources were not easy to find on the hillsides covered with forest, tall grasses, and grazing cows. Older men sought to optimize their chances of finding ore by excluding younger men in their forays or by including only younger relatives, such as sons and nephews, whom they could easily control. Women did not have to be discouraged from participating in these ventures because of the certain danger involved, such as the collapse of a mine. Moreover, they could not be away from their domestic responsibilities for days at a time.

The search team looked for a dung beetle, an *ekijunjumira,* whose hind legs dug through cow and elephant dung into the soil below and, in the process, sometimes brought up glittery black stones (Childs 1998). When and if other members of the search team verified that these stones were ore, the man who first found the beetle and its leavings proclaimed himself the discoverer and owner of the hill of ore. It is likely that this claim was sometimes contested.[7] Whoever won the discovery immediately called on the *nyakatagara* (spirit medium) of his *omuchwezi* (family deity). The word of the discovery traveled rapidly, and many people hastened to gather at the new mine with the *nyakatagara.*

This social gathering had three key functions. First and most important, it was the context in which the discoverer's rights were made known and publicly accepted, and he acknowledged his obligatory expenses. Second, it was the context for the first necessary ritual associated with, and critical to, the entire iron-making sequence. Third, it was a celebration of the future wealth that the discoverer, his family, and associates would reap. The *nyakatagara* began the ritual by making contact with the discoverer's family spirit and then slaughtering a white sheep that the discoverer was obligated to provide. She immediately dripped the sheep's blood into the mine pit and then spat on the people around her using water or beer, also provided by the discoverer. The sheep was then roasted, and everyone feasted and drank. The discoverer compensated the spirit medium for her effort with a piece of iron.

These actions were performed to achieve a number of positive outcomes: the family deity was thanked for the discovery of the mine, and the rituals helped "to bewitch any evil that would spoil the mine," such as a mine shaft collapse or the death of a miner.[8] The acts also helped ensure that the ore would not spoil during smelting. Finally, luck and fortune was extended to all the people gathered, including the discoverer. The use of a white sheep in this context was deliberate, in order to generate and reinforce particular symbolic meaning. White meant—and still means—purity and luck among the Toro and neighboring Nyoro (Beattie 1968b; Needham 1967).[9] Because of its meaning and significance, white was used at other ritual and technical occasions associated with ironworking.[10]

The other men and women who came to the event at the new hill of ore were usually members of the discoverer's family and clan. As well, men, often ironworkers from other clans and villages who had undergone a blood brotherhood ritual with the discoverer, also attended. Women, particularly the wives of the search team and the miners, were allowed to participate in the ritual and feasting unless they were menstruating. That condition, the ironworkers made clear, would "spoil" the ore when it was being smelted.

In light of the social rules and proscriptions on attendance, most of those who chose to participate in these events were being opportunistic. They knew that the celebration involved a feast of roasted meat. Probably of more importance were the benefits they hoped to reap from the good fortune of the discoverer, through the potential jobs he had to fill and the wealth he might distribute. Receiving the blessing of the family spirit through the *nyakatagara* was also important to all attendees.

Serious mining began the day after the ritual and celebratory feast. Given the hard work involved and the dangers associated with digging deep pits, young men who participated in the inaugural ritual and feast were often enlisted as *abahaige* (miners). They used an iron hoe to dig out the top layers of soil and an *omusuma* (iron pick) to mine the ore in the shaft, both tools usually provided by the discoverer. The *fundi,* who sorted out the ore at the edge of the mine pit, tended to be older smiths with firsthand knowledge of the physical properties of the ore. Both sets of workers negotiated their payment with the discoverer before they began work. All knew that, as part of the job, they had to abstain from sexual relations or the ore would fail during the later phase of smelting. The group retrieved enough ore to conduct the number of smelts immediately planned by the discoverer.

Family members, clansmen, and neighboring ironworkers often asked the discoverer's permission to open other mine shafts in the hill he now owned. As he assigned them places to work, these mining groups negotiated an appropriate compensation for this right, usually based on the amount of ore they would recover. They paid the discoverer either in ore or with pieces of iron that they later smelted. No rituals were performed as each new shaft was dug on a hill.

A second kind of iron ore was also collected as an iron smelt was planned. This was a dense, red clay-like material called *entabo.*[11] Evidence to date indicates that there was no ritual associated with the discovery of *entabo.* Once acquired, however, it was combined with the black, glittery ore and left to dry together. This was done to "befriend" or "have sexual intercourse with" the black ore so that they would stick together during smelting. Although not mentioned by my Toro informants, neighboring Nyoro ironworkers have reportedly referred to the hard, black ore and the soft, red ore as male and female, respectively (Roscoe 1923).

Smelting

As soon as ore was found, the lucky man sent word to his most trusted wife (often he had four or five): "Do not untie your belt."[12] This meant that she must be faithful to her husband until he returned and be careful about who came to greet her. When the discoverer brought the two types of ore to dry in his compound, a broader group of rules directly applicable to his family were immediately set in motion. Most important, he and his most trusted wife now engaged in sexual relations until the ore was ready to smelt. "Immediately after the ore was discovered, you [wife] started embracing each other in bed. You never turned away from your husband."[13] Perhaps this was done to reinforce, mimic, and play out a technically and ideologically based activity that was occurring as the ores dried, or would occur in and involve the furnace itself. As well, neither husband nor wife could cheat on each other during this time. No one could touch the wife, particularly on her shoulder, which meant bad luck, or around her stomach, which was disrespectful and suggested infidelity.[14] The only people who could touch the ore as it dried were the discoverer, his trusted wife, and small children who were considered to be "pure." The most important and fundamental social relationships in operation during this period of protecting the ore, then, were those that stressed fertility and marital fidelity.

The black and red ores were mixed during the final days of drying, a process that might take up to a week. When the ore was ready, a smelt was prepared by the discoverer or a fellow clansman. Only experienced men smelted; usually two or three individuals spelled each other on the bellows. Also important, none of them could sleep with their wives the night before they smelted.

The discoverer and his fellow workers dug a furnace pit approximately 3 feet in diameter and 2 to 3 feet deep and built a small shelter over it to protect the bellowers from the sun. They placed a pair of bellows at one side of the pit and aligned the flared mouth of the *tuyere* with the two nozzles of the bellows.[15] They set the length of the *tuyere* into the pit so that it lay above layers of ash, reeds, and some charcoal and below layers of charcoal and ore. Notably, they did not build a superstructure or walls above the pit, as was done by ironworkers in neighboring societies such as the Ganda (Roscoe 1911), the Haya (Schmidt and Avery 1978; Schmidt and Childs 1995), and by the royal smelters of the Nyoro (Roscoe 1923). Not only did these shaft furnaces usually smelt a greater volume of ore and yield more iron bloom than the Toro furnaces, but the walls also provided a possible medium to express other, nontechnical aspects of the technology (Childs 1991; Herbert 1993), such as those discussed below.

As the furnace was being prepared and then lighted to begin the smelt, the discoverer and owner of the ore asked the *nyakatagara* to come and make contact with the family spirit. She brought her buffalo horn (*ihembe*) and other necessary ritual gear (*emyarro*) to facilitate the process. The owner of the ore, or in some cases an individual who wanted a sizable piece of iron from the smelt, supplied a white sheep for slaughter. To ensure bountiful success, the *nyakatagara* sprinkled the sheep's blood into the furnace (*ejugutiro*), on the bellows, and on the ironworkers. Others, mostly older men, deemed to be clean and not potentially threatening to the process, could watch the proceeding and partake of the roasted meat.

During these furnace preparations and the actual smelting activities, certain other people were not allowed nearby for fear that their presence would cause the ore to turn black in the furnace and "die."[16] For example, one of the smelter's wives often brought food for the workers but had to call from a distance to announce her arrival. Neither she nor any woman, except for the *nyakatagara,* were allowed at the site. Those who were menstruating were considered particularly dangerous to the furnace and its success. No one, man or woman, who had been unfaithful to their spouse could look at the furnace or the ore. And no one, including the smelters themselves, were permitted by the furnace if they could not control their flatulence, since it was feared that the gas would enter the ore and kill it. Again, fidelity and the exclusion of women who failed to conceive structured the very nature of the social interactions at a Toro smelt, while fundamental notions of life and death influenced the practical activities of iron production.

In order to ensure that passers-by followed these rules, the smelters set up an *enkomo* (blockade) across the path to warn people of their activities. Some Toro ironworkers constructed a fence around the smelting space, using a particular plant, the *orusororo,* whose derivative verb translates to "separate" or "pick and choose" (Davis 1938:164). As well, the person in charge, such as the owner of the ore, informed any onlookers of the rules and appropriate behaviors.

Many of the social values related to these technical procedures were represented and reinforced symbolically during the smelting process. Whereas other ethnic groups of ironworkers used their furnace walls to express particular beliefs or social relationships related to ironworking, Toro men utilized certain iron-smelting tools in this way. The pair of clay bellows pots, for example, signaled the

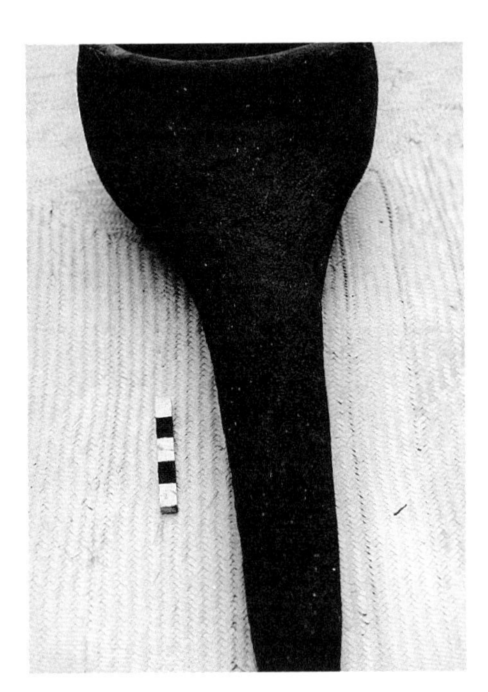

Figure 2.2. The roulette decoration on the female bellows of the Ndunga family.

reproductive activity that was soon to take place in the furnace. One bellows pot was female, called *nyinabarongo,* the mother of twins, and the other was male, called *esebarongo,* the father of twins.

The female bellows was signified by a roulette design all around the pot (figure 2.2), a typical decoration on various kinds of Toro pottery. Notably, the bellows potters from Nyoro and other neighboring polities molded representations of distinctly female genitalia on the pot (Lanning 1954). Similarly, the Toro male bellows was constructed with a representation of a penis located above the air shaft, which extended into the *tuyere.*[17] Information gleaned from my interviews, and those of Lanning (1954) with nearby ironworking groups, indicate some stylistic variation in how bellows pots were decorated. Stylistic diversity was probably due to a number of factors: the varying skills of individual potters, group differences in the degree of explicitness required to indicate the gender of the bellows, and peer pressure related to ideological changes in these societies under colonial influence.

These two anthropomorphized bellows[18] were tied together under one sheepskin, with the male to the left and the female to the right. This is the customary spatial arrangement of Toro men and women in many social situations, including sleeping in the same bed. A stick was then tied into the sheepskin above each bellows pot. The bellower rapidly lifted and plunged each stick in an alternating rhythm to efficiently push air into the furnace during smelting.

The sexual pairing of the male and female bellows combined with the efforts and comportment of the sexually potent and faithful ironworkers yielded great wealth—an iron bloom referred to as *abarongo,* or twins. Ndunga explicitly noted that bellowing and smelting were activities of creation. The birth of twins, however, is a mixed blessing among the Toro and the neighboring Nyoro (Beattie 1962) because it connotes not only abundance and prosperity, but also serious ritual danger that might result in death. The parents of twins must initiate and perform expensive ritual obligations with the help of the family *nyakatagara* to counter this ritual danger. In the end, they are nonetheless held in very high esteem (Beattie 1962:1). The paired clay bellows, then, not only effected a practical end—fueling a furnace fire with requisite oxygen—they simultaneously reinforced and materialized fundamental notions of life, death, and fertility.

Furthermore, if a wife of an ironworker gave birth to a girl on the day he smelted, the baby was given a special name—"Kabahesi." This roughly translates to "smith's girl." According to Toro beliefs, she was destined to marry a smith, be fertile, and produce many children. In contrast, a baby boy born at that time was not named in any special way.

The end of the smelt was a time of celebration called the "serving," when old men and women, "good" young men, and women who were not menstruating could participate. This event took place when the bloom was pulled out of the furnace by two to four *abahereza* (servers) with long sticks called *emihano.* The bloom was then hammered out into a rough shape and split into a number of paired pieces with an axe by the *omwasi w'ebyoma* (splitter). It was at the "serving" that all individuals (primarily family and clanspeople) involved in the various tasks leading up to the smelt—such as miners, ore separators, charcoal makers, bellows makers, bellowers, and servers—were compensated for their services with pieces of iron. They were obliged to attend the "serving" to ensure their rightful share, although the discoverer and owner of the ore had the final say on the size of the appropriate piece for each person.

There was singing and dancing at this celebration. The songs to the family spirit were so powerful that young, unmarried girls who might be "climbed" or entered by the spirit were strongly discouraged from attending. Some young girls did take the opportunity, however, because to be "climbed" was the first step in becoming a *nyakatagara.* This specialist role in Toro society was one of the very few ways in which women could attain some degree of wealth, power, and status within their community (Beattie 1961, 1968a). Not only was the *nyakatagara* highly respected, but she was well paid to perform rituals on occasions that needed sanction or extra help from the family spirit. These circumstances included difficult or inexplicable times, such as a miscarriage, the birth of twins, or a strange death; complex economic activities, such as hunting or ironworking; and, occasions when a person was "climbed" by the spirit and needed to be instructed into spirit mediumship.

Forging

The final phase in the production of an iron object was forging. It was more of a year-round job than either mining or smelting, which were necessarily interrupted by seasons of damaging rainfall and the demands of other subsistence activities.

The forge was—and still is—a small, roofed hut open at the sides. It contained a forge pit with its pair of gendered bellows, a pit of water to douse the flames, at least one anvil, and various hammers and other tools (figure 2.3). Also, unlike the mine and the smelting furnace, which were restricted spaces, the forge was—and still is—a social place. "It looked like a traditional market," since anyone could come by the forge to buy, trade, chat, or pass the time.[19] All passersby,

Figure 2.3. The forge of smith Bamulimbya. Visible are the pair of clay bellows under one sheepskin, the pot of water and grass bundle (*isiza*) for dowsing flames, and the anvil (foreground).

including women and the king of Toro, were encouraged to pump the bellows. Even daughters of the smith sometimes helped at the bellows when no boy or man was available. They were required to kneel as they pumped, however, because girls were not permitted to take the typical position of squatting. It is likely that this rule of bodily comportment was required because in a squatting position the bellows and the iron being forged would be exposed to female genitalia, thus exposing the forge to threat.

Each forge was built, operated, and maintained by an *omuhesi omukuru* (master smith). He had a number of rights and responsibilities, including training his apprentices and assistants, supplying the forge with essential raw materials such as charcoal and raw iron, and properly storing the tools. He also weighed and arbitrated any interpersonal differences experienced at the forge, over the price of a forged object, for example, or the theft of a piece of iron.

Since the craft of ironworking was a lucrative one among the Toro and the master smiths were usually men of high social status and some wealth, they also had to protect their workplace and interests from any potential evil, ill will, and danger. These factors might become manifest as inexplicable difficulties involved in forming or working iron, enigmatic events at the forge that might suggest evil or sorcery, and atypical illnesses experienced by the smith's assistants or members of his family. The master smith generally resolved such problems by going to the family *nyakatagara* for communication with his family *omuchwezi*. The master smith also took several other precautions. For example, he healed (*kutamba*) the forge in the morning before he began work with a cleansing process during which he employed the *isiza* (a grass sprinkler normally used to stifle flames during forging) to sprinkle water around the forge and on the tools. He also was careful to avoid having intercourse with any woman the night before he forged. Instead, he "slept on one side."[20] In addition, he sacrificed a sheep to the forge and to his *omuchwezi* at least once a year.

Ndunga told me that the guidance and blessing of his family spirit, *Omuchwezi Ntogota,* was sufficient to keep evil and sorcery away from the forges used by himself, his father, and his grandfather. His family did not need to use special medicines, as other African societies commonly do to protect the ironworking spaces (Childs and Killick 1993; Herbert 1993). After many conversations, however, I learned that the ironworkers in his *Abachwamba* clan wore the hide of a sheep at their mines, their smelts, and their forges. Sheepskin had special properties of protection that, in combination with the proper rituals to *Ntogota,* helped ensure success and prosperity.

The most significant event in the economic life of a smith was when he first set up his own forge, called the *nguraho* (opening). This occurred upon the death or retirement of the mentor who taught him his skills, usually his father or uncle, or upon his departure from the family forge to take up work in a village that did

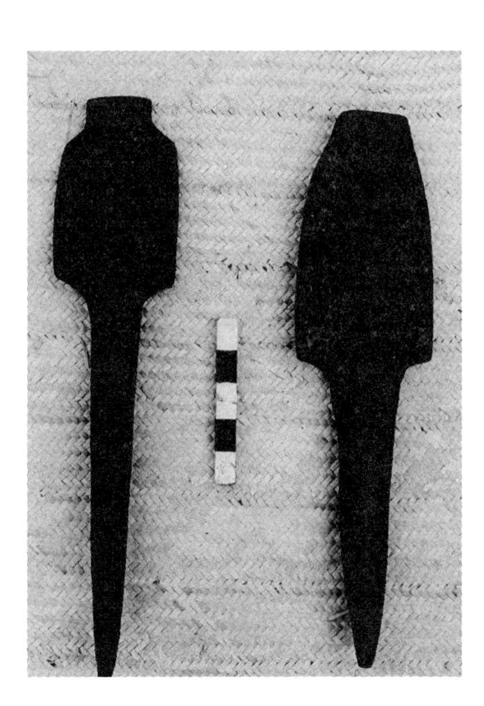

Figure 2.4. Two hammers of the Ndunga family. The one on the right is called the "mother," the one on the left is considered one of her children.

not have a smith. The *nguraho* began with the gathering of a group of men, usually immediate family and clansmen, who then went into the hills to find and prepare an anvil stone. As the group carried the stone back to the site of the new forge, they sang the songs of ironworkers. Many people gathered at the new forge where, except for the exclusion of young girls, there were no social restrictions. As at the "serving" of the smelted iron, the fear was that young girls might become possessed by the family spirit while the powerful songs were being sung.

Before the anvil was put into its new position, the *nyakatagara* for the spirit of the new smith's family was summoned. The smith then slaughtered a white chicken or sheep and the *nyakatagara* dripped its blood into the pit where the anvil would lie. She also sprinkled its blood on those who helped carry the anvil to the forge. The hammers, particularly the male *omurubiko* (a large, round stone), the female *enyondo* (figure 2.4), and the smaller iron hammer called "the child," were brought to the anvil, where they were also "washed" with blood by the *nyakatagara*.

Next, the *nyakatagara* handed the new smith his hammers. His father or another smith then spoke about the virtues of being a good smith, especially the importance of kindness and help to all in times of need:

> My child, I've handed you this hammer. If someone is hungry and wants a hoe, or his hoe is broken, never refuse to help him. I've handed you this hammer, forged the king's spears and basket needles.

> I hand you this hammer. Forge for the deformed/lowest class and the respected. Whatever they bring for you to forge, you forge. Never be angry. A smith never has anger. Don't shout at someone when he brings something or bark at him and tell him "Take your things."[21]

This senior smith also extolled the significance of the hammer and the wealth it would bring to the smith, his family, and his community. These words emphasized and reaffirmed the ways in which the work and very being of the smith was part of a complex web of social principles, values, and material symbols. The new smith then immediately forged two objects with his new hammers and anvil. One went to the *nyakatagara* for her services and the blessings she transmitted from the family spirit; the other went to the smith's mother for her part in making his success possible. The ceremony ended with more celebrations, during which the hammer—as the symbol of wealth and power for a knowledgeable few—was the focus of several songs and dances (Childs 1998).

DISCUSSION

Toro ironworking was as much about reinforcing and reaffirming fundamental social relationships as it was about making iron. Elder, master ironworkers combined technical contexts with ritual ones at the discovery of a new source of iron, the iron smelting furnace, and the forge to play out and materialize social principles and values embodied in such dichotomous relationships as husband/wife, parent/child, barren/fertile, elder/youth, living/dead, and rich/poor. This was done to gain and maintain prestige, wealth, and social influence. The wealth attained through discovering, collecting, or trading iron ore, and through making iron bloom and forging objects, allowed experienced ironworkers to have several wives, many children, and influence over a broad network of people. "After all, a hoe bought a wife."

Although these ironworkers placed considerable emphasis on following specific social rules and cultural principles to further those ends, they also tolerated a notable degree of negotiation, contest, and opportunism. Family members, clanspeople, blood brothers, and the *nyakatagara* actively benefited from their social relationships to feast at ironworkers' celebrations, to claim jobs and negotiate compensation in the ironworking process, and to receive the blessing of the family spirit. A few young girls, typically excluded from these activities, sometimes took advantage of such contexts so that they might be entered or "climbed" by the spirit and, through this experience, eventually become a spirit medium.

The multiple uses of technical contexts merit careful study by archaeologists despite the attendant difficulties of recognizing ritual and symbolic space(s) in general. This is particularly difficult when an ancient community purposely destroyed or used up all possible traces of ritual activity, such as the Toro sacrifice and feasting of the sheep at the discovery of new iron ore. Through careful consideration of available ethnographies, however, archaeologists are beginning to better interpret enigmatic features and spaces in the archaeological record. For example, the presence of small holes and ceramic pots in the bottom of prehistoric iron smelting furnaces in eastern Africa are now better understood as having been aids in preventing sorcery and promoting technical success (Schmidt and Childs 1985; Schmidt and Mapunda 1996). Identifying ritual activities in the domestic spaces of an ancient society might also shed light on technological spaces in the same society and lead archaeologists to consider their possible connections.

This study also demonstrates that not one, but several types of objects used in iron production—particularly the bellows, hammers, *tuyeres,* and sheepskin—materialized the web of social relationships and values central to Toro society through their use. The mother and father bellows effectively had intercourse during the iron smelt that bore twins and the wealth they symbolized. The male hammer, a round stone used to beat out the bloom, worked in concert with the female iron hammer to produce iron objects for society. The smaller hammers, used to put the final touches on iron objects, were spoken of as children of the mother and father hammers.

All of these objects have distinguishing characteristics that allow archaeologists to differentiate them according to stylistic, technical, functional, and contextual attributes. For example, Lanning's (1954) study of the decorations placed on the male and female bellows pots among the nearby Nyoro and Ngulunga, in conjunction with my work among the Toro, reveals some variation in the types and methods of decoration, as well as the degree of abstraction used to create symbolic meaning.

Clearly, it is difficult for archaeologists to determine with some degree of certainty the technical, social, or symbolic reason(s) for stylistic variation detected through time and across space. To do so requires not only a regional approach to a study of ancient technology but also the ability to find a sizable number of related sites within the region. This approach is possible in Africa, for example, where numerous metalworking sites have been identified (Bisson 1976; de Barros 1986; van der Merwe and Scully 1971), and some scholars are beginning to examine such variation (Childs 1986; Fowler 1989; Schmidt 1997). It is also difficult for archaeologists to determine the symbolic import and meaning(s) recovered objects had in their original sociotechnical contexts. This study, however, suggests that the discovery of molded clay decorations on broken pottery sherds in an ancient iron-smelting context could permit more convincing inferences about the original vessel type and its symbolic uses. A search for blood residue on ancient iron-working tools might provide additional insights.

With increased awareness of the complex social relationships that abound in technological contexts, further refinements to the archaeology and anthropology of past technologies, including those offered in this volume, are now possible. Critical to such refinements is greater appreciation of the tendency of humans to recreate those relationships in their work settings and through the raw materials and tools they use.

NOTES

This project would not have been possible without the aid and support of a number of individuals and institutions. Peter Robertshaw provided financial assistance (NSF grant, SBR-9320392) and logistical and collegial support, for all of which I am forever grateful. John Sutton and Justin Willis of the British Institute of East Africa generously provided use of a Land Rover in both 1994 and 1995, and Paul Mutunga spared me from having to drive in 1994. Charlotte Karungi was my translator and friend during both field seasons and did a tremendous job. I wish to thank the Uganda National Council for Science and Technology for research clearance, the Ugandan Department of Antiquities and Museums, particularly Dr. Ephraim Kamuhangire, for their support, and the District Executive Secretary of Kabarole and his assistants for their assistance. Thanks are also due to several people who commented on earlier versions of this essay: Robert Ehrenreich, Eugenia Herbert, and Chris Hoffman. Finally, I owe my greatest appreciation to the many wonderful people I met and interviewed in Toro, especially those in the Butiiti area where we lived and centered our work, such as Amooti Majiri, Sergeant Jane Kajura, and the Chief of Butiiti. I dedicate this chapter to the memory of Mzee Ndunga who had unending patience with my relentless questions, was determined to get the details of precolonial and early colonial Toro ironworking recorded for Ugandans, and, along with his wife Abwooli, provided me constant hospitality.

1. Mzee N. Ndunga was born in 1914 and died in December 1995. His father taught him to be an ironworker and, as a young man, his father took him to the mines and to watch the ore being smelted in a furnace. Ndunga later became a primary school teacher for the British and supplemented his income as an ironsmith. Other informants for this study include Mzee Bamulimbya, a retired smith, approximately 75 years old; Z. Kirigwajo, a retired smith, approximately 80 years old; P. Tinkasimere, a practicing smith, 68 years old; A. Mpeerabusa, a retired smith, 80 years old; and P. Bamanyisa, an intermittent smith and a diviner, approximately 40 years old. Ndunga first mentioned the relationship between a wife and a hoe during conversations in July 1994. Kirigwajo and Mpeerabusa made similar remarks on other occasions.
2. Stated in an interview with P. Tinkasimere, July 14, 1995.
3. The relationships between a hoe, cow, and wife was nicely summarized by Ndunga on July 28, 1994, when he said: "A hoe was very profitable. It is the one that married a

wife. It would marry the wife—only one hoe. And in a hoe, someone could get a calf. If someone had many cows, he would get one calf and use it to buy a hoe. All those are valuable."

4. The prefix "omu" refers to a single person, while the prefix "aba" refers to more than one person in LuToro, the local language.
5. I can only estimate the size of an average bloom from my informants' descriptions, since no one still owned one. Blooms that existed when the Toro king and British outlawed mining and smelting in the late 1920s were either quickly forged into an object, or buried, or hidden elsewhere for safekeeping. It is likely, however, that blooms averaged 5 to 10 kilograms.
6. P. Bamanyisa, July 19, 1995.
7. My informants did not offer any examples of contests between team members over who discovered ore although they certainly implied this happened. I believe that my informants had no direct experiences to relate because they were relatively young when mining was still practiced.
8. N. Ndunga, July 25, 1994.
9. In the 1830s, a group of people broke away from their homeland in Nyoro and called themselves Toro. A relatively recent history of the Toro is provided by Ingham (1975) who gives some detail of the separation between the Nyoro and the Toro. It is important to note that many aspects of the language, cultural values, and social system of the two groups do not differ substantially. It is possible, therefore, to use historic and ethnographic information on the Nyoro to gain insights into the activities and belief systems of the precolonial and early colonial Toro and to use dictionaries of LuNyoro for translations (Davis 1938).
10. The iron produced in a successful smelt is considered to be "white." There is a Toro smithing song that refers to the hammer as "white" (Childs 1998).
11. Analyses to identify the chemical compositions and physical characteristics of the two types of ores are being arranged.
12. N. Ndunga and A. Kamara, July 15, 1995.
13. N. Ndunga, July 22, 1994.
14. If a man puts his hands on a woman's shoulders, it can symbolize having intercourse (Beattie 1962:9). The "stomach" is an euphemism for her genitals.
15. The Toro phrase for smelting is *kujuguta obutale.* This roughly translates to "bellow/blow air on ore," which emphasizes the importance of the bellows to the whole smelting process.
16. In other words, the usual signs of a successful smelt—such as the red and white colors of hot ingredients, the hissing sounds, and the sights of dripping slag—would not be experienced if someone looked through the *tuyere* hole.
17. Although several elder Toro smiths described the male bellows pot in great detail and recognized its importance to the smelting process, I never saw one in the field. This is because no one I interviewed still owned one with the explicit imagery, a practice that had been profoundly discouraged by Christian missionaries in the area since the early 1900s. The Uganda National Museum, however, has several examples of Nyoro male bellows pots.

18. If one of the bellows pots cracked, it was said to be ill. If a bellows broke, it was said to *kuzara*, "to bear, give birth to" (Davis 1938:177).
19. N. Ndunga, July 19, 1994.
20. P. Bamanyisa, July 19, 1995.
21. N. Ndunga, July 26, 1994.

REFERENCES CITED

Beattie, J.

1961 Group Aspects of the Nyoro Spirit Mediumship Cult. *Rhodes-Livingstone Institute Journal* 30:11–38.

1962 Twin Ceremonies in Bunyoro. *Journal of the Royal Anthropological Institute of Great Britain and Ireland* 92:1–12.

1968a Spirit Mediumship in Bunyoro. In *Spirit Mediumship and Society in Africa,* edited by J. Beattie and J. Middleton, pp. 159–170. Africana Publishing Corporation, New York.

1968b Aspects of Nyoro Symbolism. *Africa* 38(4):413–442.

1971 *The Nyoro State.* Clarendon Press, Oxford.

Bisson, M. S.

1976 *The Prehistoric Coppermines of Zambia.* Ph.D. dissertation, Department of Anthropology, University of California, Santa Barbara. University Microfilms, Ann Arbor.

Childs, S. T.

1986 *Style in Technology: A View of African Early Iron Age Iron Smelting through Its Refractory Ceramics.* Ph.D. dissertation, Department of Anthropology, Boston University, Boston. University Microfilms, Ann Arbor.

1991 Style, Technology, and Iron-smelting Furnaces in Bantu-Speaking Africa. *Journal of Anthropological Archaeology* 10:332–359.

1994 Society, Culture and Technology in Africa: An Introduction. In *Society, Culture, and Technology in Africa,* edited by S. T. Childs, Supplement to Vol. 11, pp. 6–14. MASCA Research Papers in Science and Archaeology, Philadelphia.

1996 Technological History and Culture in Western Tanzania. In *Culture and Technology of African Iron Production,* edited by P. R. Schmidt, pp. 277–320. University of Florida Press, Gainesville.

1998 "Find the *Ekijunjumira*": Iron Mine Discovery, Ownership, and Power among the Toro of Uganda. In *Social Approaches to an Industrial Past: The Archaeology and Anthropology of Metalliferous Mining,* edited by A. B. Knapp, V. C. Pigott, and E. W. Herbert, pp. 123–137. Routledge, London.

Childs, S. T., and D. J. Killick

1993 Indigenous African Metallurgy: Nature and Culture. *Annual Review of Anthropology* 22:317–337.

Cline, W. B.
1937 *Mining and Metallurgy in Negro Africa*. General Series in Anthropology No. 5. George Banta, Menasha, Wisc.

Conkey, M. W.
1990 Experimenting with Style in Archaeology: Some Historical and Theoretical Issues. In *The Uses of Style in Archaeology*, edited by M. W. Conkey and C. Hastorf, pp. 5–17. Cambridge University Press, Cambridge.

David, N. C., J. Sterner, and K. Gavua
1988 Why Are Pots Decorated? *Current Anthropology* 29:365–389.

Davis, M. B.
1938 *A Lunyoro-Lunyankole-English and English-Lunyoro-Lunyankole Dictionary*. Uganda Book Shop, Kampala.

de Barros, P. L. F.
1986 Bassar: A Quantified, Chronologically Controlled, Regional Approach to a Traditional Iron Production Centre in West Africa. *Africa* 56:148–174.

Dobres, M-A.
1995 Gender and Prehistoric Technology: On the Social Agency of Technical Strategies. *World Archaeology* 27(1):25–49.

Dobres, M-A., and C. R. Hoffman
1994 Social Agency and the Dynamics of Prehistoric Technology. *Journal of Archaeological Method and Theory* 1(3):211–258.

Fowler, I.
1989 *Babungo: A Study of Iron Production, Trade and Power in a Nineteenth-Century Ndop Plain Chiefdom*. Unpublished Ph.D. dissertation, University of London, London.

Gosselain, O. P.
1992 Technology and Style: Potters and Pottery among Bafia of Cameroon. *Man* 27:559–586.

Herbert, E. W.
1993 *Iron, Gender, and Power: Rituals of Transformation in African Societies*. Indiana University Press, Bloomington.
1998 Mining as Microcosm in Precolonial Sub-Saharan Africa: An Overview. In *Social Approaches to an Industrial Past: The Archaeology and Anthropology of Metalliferous Mining*, edited by A. B. Knapp, V. C. Pigott, and E. W. Herbert, pp. 138–154. Routledge, London.

Hodder, I.
1986 *Reading the Past*. Cambridge University Press, Cambridge.

Ingham, K.
1975 *The Kingdom of Toro in Uganda*. Methuen & Company, London.

Ingold, T.
1991 Society, Nature, and the Concept of Technology. *Archaeological Review from Cambridge* 9(1):5–17.

Kense, F. J.

1983 *Traditional African Iron Working.* African Occasional Papers No. 1. Department of Archaeology, University of Calgary, Calgary, Alberta.

Lanning, E. C.

1954 Genital Symbols on Smith's Bellows in Uganda. *Man* 54(262):167–169.

Lechtman, H.

1977 Style in Technology: Some Early Thoughts. In *Material Culture: Styles, Organization, and Dynamics of Technology,* edited by H. Lechtman and R. S. Merrill, pp. 3–20. West Publishing Co., St. Paul, Minn.

Miller, D., and N. J. van der Merwe

1994 Early Metal Working in Sub-Saharan Africa: A Review of Recent Research. *Journal of African History* 35:1–36.

Needham, R.

1967 Left and Right in Nyoro Symbolic Classification. *Africa* 37:425–452.

Pfaffenberger, B.

1988 Fetished Objects and Humanised Nature: Towards an Anthropology of Technology. *Man* 23:236–252

Roscoe, J.

1911 *The Baganda.* Cambridge University Press, Cambridge.

1923 *The Bakitara.* Cambridge University Press, Cambridge.

Schiffer, M. B.

1992 *Technological Perspectives on Behavioral Change.* University of Arizona Press, Tucson.

Schmidt, P. R.

1983 An Alternative to a Strictly Materialist Perspective: A Review of Historical Archaeology, Ethnoarchaeology, and Symbolic Approaches in African Archaeology. *American Antiquity* 48:62–79.

1996 *Culture and Technology of African Iron Production.* University of Florida Press, Gainesville.

1997 *Iron Technology in East Africa. Symbolism, Science, and Archaeology.* Indiana University Press, Bloomington.

Schmidt, P. R., and D. Avery

1978 Complex Iron Smelting and Prehistoric Culture in Tanzania. *Science* 201:1085–1089.

Schmidt, P. R., and S. T. Childs

1985 Innovation and Industry during the Early Iron Age in East Africa: The KM2 and KM3 Sites of Northwest Tanzania. *African Archaeological Review* 3:53–94.

1995 Ancient African Iron Production. *American Scientist* 83(6):524–533.

Schmidt, P. R., and B. B. Mapunda

1996 Ideology and the Archaeological Record in Africa: Interpreting Symbolism in Iron Smelting Technology. *Journal of Anthropological Archaeology* 16:73–102.

Taylor, B. K.

1962 *The Western Lacustrine Bantu.* International African Institute, London.

Tylecote, R. F.

1986 *The Prehistory of Metallurgy in the British Isles.* The Institute of Metals, London.

van der Merwe, N. J.

1980 The Advent of Iron in Africa. In *The Coming of the Age of Iron,* edited by T. Wertime and J. Mulhy, pp. 463–506. Yale University Press, New Haven, Conn.

van der Merwe, N. J., and R. T. K. Scully

1971 The Phalaborwa Story: Archaeological and Ethnographic Investigation of a South African Iron Age Group. *World Archaeology* 3:178–196.

3. Crafts and the Evolution of Complex Societies: New Methodologies for Modeling the Organization of Production, a Harappan Example

VALENTINE ROUX AND PIERRE MATARASSO

A TECHNOLOGICAL APPROACH IS A PRIVILEGED ONE for understanding the functioning and evolution of cultural formations. In an attempt to reconstruct technosystems, it presupposes a complex number of relationships among the individuals participating: relationships that are symbolic, economic, hierarchical, and social. These relations are closely interwoven with the technoeconomic system or material culture (Lemonnier 1986, 1993; see the studies in this volume). It follows that a picture of the "materiality" of a technosystem permits the reconstruction of the social relations involved.

Such a picture can be obtained by identifying and quantifying material technical practices, which in turn raise questions about the social organization of the productive system. Given the argument that craft production is constituted by technical operations (in the sense of the *chaîne opératoire;* see also Dobres chapter 6 this volume), how do the different operations fit together?[1] Were they achieved by a single worker or by groups of workers? If a single worker, did this individual perform the technical operations to produce an object in sequence from beginning to end, or on the contrary did this person practice the mass production of objects one stage at a time? In the case of craftspeople working together at a workshop, were they brought to that place in their capacity as specialists to pursue all stages of the manufacture process, or did different workshops and their complement of specialists engage in distinctly different technical operations? In either case, what was the relative importance of different tasks or different specializations?

In this study, we reconstruct the technosystem of Harappan stone beads. Our goal is, first, to ascertain from the number of beads found at Harappan sites something about annual bead production and the number of workers involved, and on this basis to ask how "craft specialists" organized bead production, what status bead makers had in Harappan society, and what "function" the beads served. Hypotheses can then be formulated to suggest which factors may have played a role in the formation of the Indus civilization.[2]

The Harappan technosystem can be reconstructed by interpreting archaeological data in light of quantified ethnoarchaeological observations ascertained through the method known as *activity analysis* (described below). This method makes it possible not only to measure systems of production but also to envision alternative forms of organization whose plausibility may be assessed in terms of local conditions (the archaeological context).

METHODOLOGY

To ascertain the number of technicians at work and their technoeconomic organization from the known number of finished products, one must look to the technical practices involved, but beyond them as well. An additional referent is required wherein technical activities have been measured from the point of view of consumption (in terms of raw material, energy, duration of work, and the like) and production (number of objects made per day). Constructing such a referent presupposes the existence of a living traditional technoeconomic system in which technical practices are analogous to the ones employed in the past.

Appeal to Ethnoarchaeology

The chain of fabrication techniques used in Harappan stone bead production has been identified on the basis of "diagnostic" criteria (Gwinett and Gorelick 1981; Kenoyer 1986; Mackay 1937; Pelegrin 1994; Piperno 1973; Roux 1999; Tosi and Piperno 1973; Vidale 1986). This diagnostic fabrication chain involves the following technical operations (and diagnostic tools): heating, knapping (percussive tools), grinding (grinding stone), drilling (drill bow), polishing and shining (grinding stone, leather bag, polisher bow). This sequence of operations is directly observable in its totality at contemporary workshops producing stone beads in Khambhat (India) and in part at some workshops producing carnelian bezels in Yemen. In India, beads are produced at a monumental scale and by various technical modes. Bead production in Yemen, though more anecdotal, yields information on grinding, polishing, and shining operations that are now mechanized in India.

In this study, elementary technical operations are considered cultural invariants and the quantitative data obtained about them, cross-cultural. That is, they apply to all technical activities present or past, given the level of the technological analysis. This level remains low and deliberately ignores the specifics of the cultural context in which the technical activities take place. Thus our ethnoarchaeological study takes into account determinant factors relating to the objective properties of material resources, environment, and subjects, for example, the possibilities inherent in the raw material itself (that is, given core stones of certain dimensions and utilizing a particular method and technique of manufacture, how many beads can an individual produce?). The measures that will have to be adjusted to archaeological situations will concern, for example, the amount of time it takes to knapp beads using indirect percussion by rebound, which is surprisingly shorter than all other known knapping techniques (including percussion and pressure).

A Method Derived from Industrial Economy: Activity Analysis

The work of von Neumann (1945), Dantzig (1963), and Koopmans (1951) has developed into a formal method for describing production and economy, known as "activity analysis." The method has been applied in a number of case studies in practical domains such as industrial economy and operational research, in the field of macroeconomics, and in theoretical models of general equilibrium. *The object of activity analysis is to determine at the most general level possible the constraints and alternatives underlying the organization of complex systems of production.* Such an analysis has the potential to provide information on both the organization of a living system and any number of possible alternative forms for its organization. The constraints of primary concern to activity analysis are those at the link between the material nature of resources and their disposal by producers, that is, whether production is being strongly constrained by natural resources or, conversely, whether surplus production can be, and is, generated. From this perspective, activity analysis is compatible with an "economicist" view of optimization in the production of limited resources as much as it is with an "anti-utilitarian" view of the redistribution of surplus, sensu Bataille or Sahlins.

The formalized method introduced by von Neumann was designed to break down a complex productive process into its constituents, called *activities.* An activity is an elementary technical act that can be characterized on the basis of its input and output (consumption and production). Von Neumann's concept of an activity can be exemplified with a cooking recipe. A recipe calls for a set of ingredients (flour, eggs, milk, etc.), certain instruments (oven, whisk, and so on), and labor and energy. An activity corresponds to the specifics of the recipe. At this analytic scale, a cooking recipe constitutes an activity, but it may also be described as a complex, sequentialized process formed by an ensemble of elementary activities, each with possible alternatives (for example, beating the egg mixture by hand instead of with a whisk).

At higher analytic levels, activity analysis is concerned with more complex systems that correspond to coherent associated activities. An example would be an agricultural system in a limited geographic area where labor and materials are also limited. The farmer's concerns might be how to provide adequate supplies of food for the family, or how to reap the maximum benefit by selling products outside the household. Either of these objectives could be accomplished through different sets of agricultural activities. To satisfy the nutritional needs of the family or market demands, the farmer might cultivate tubers or cereals; or the farmer might decide to plant these two cultigens because of taste, or to diversify risks against unknown factors such as the climate. The farmer might also choose to raise certain varieties of animal resources (breeding cows, goats, and others). In each case, these productive activities will be in competition (say, for land, labor, tools, and laborers). Competitive factors constitute the principal object of constraints. These are also complementary factors among these activities (for example, animals will be a source of fertilizer as well as a form of labor useful in cultivating plant foods). Complementary factors structure the alternatives to a productive system. *Activity analysis highlights the constraints and alternatives within a formal framework designed to provide quantifiable results.*

It is in this spirit of understanding constraints and alternatives that we examine bead production in Khambhat, defined here as a technosystem (analogous to an agrosystem). We focus on the complementary factors existing among major technical classes of bead production and not on micro-organizational factors constituting small-scale craft enterprises. This perspective, at least in its initial exploration here, is best suited to the archaeological constraints that we typically confront, which (more often than not) do not permit detailed consideration of past socioeconomic structures (but see such an attempt in Dobres chapter 6 in this volume). At the same time, we underscore that the analysis of microorganizations is not separable from those competitive and complementary factors structuring the overall coherency of the system. Microorganizational issues could complement a study of the Khambhat production system. What is most favorable for describing a technosystem are certain characteristics of its structure: individual technical agents or the small-scale enterprises may have complementary functions (that is, separate workshops for grinding, drilling, and polishing), or they may be integrated into a coherent technical plan (a single workshop where all the tasks are accomplished). In the first situation, it is logical to take an overall view of the system, while in the second context one might prefer to model each task separately. In reality, however, it is never simply one situation or the other. Therefore we combine the two by modeling a global system of complementary technical activities.

If a picture of a technosystem is to be as complete as possible, its observations must come from diverse perspectives. Some observations can be obtained at a local scale (such as the time it takes to fabricate one bead). Other factors can be drawn from observations of the overall ensemble at a given point (for example, the total number of artisans in a village required to perform a specific task). The last category of factors to observe concerns information about the overall system (such as the relative proportion of different beads produced in one village; this sort of information is generally more accessible than information on the overall system). The final goal of activity analysis is to integrate these different forms and scales of information into a quantitative representation in order to examine their degree of coherency.

The idea of creating an inventory of elementary technical acts in a craft, industrial, or agricultural setting is not new. Technical actors themselves have made such attempts, as well as those who study them (for example, Haudricourt 1987; Leroi-Gourhan 1943). What makes activity analysis new is the possibility of establishing a quantitative base of data that can be analyzed in order to develop a coherent representation of a technosystem. This global working of the data is only possible through relatively elaborate data processing.

Another point to note is that activity analysis would seem to be all the more appropriate for archaeological problems when conceptually attached to the notion of *chaîne opératoire,* as developed by Leroi-Gourhan and as employed today by prehistorians and ethnographers of technology (discussion in Dobres and Hoffman 1994; in this volume see Dobres chapter 6, Wake chapter 9).

ETHNOARCHAEOLOGICAL OBSERVATIONS

The first step is to assemble and organize quantitative information that will permit us to describe each of the elementary constituents of a technical system. These elements are mainly observable technical operations. Before quantitative information can be organized, it is necessary to determine, a priori, the scale at which the constituents are to be aggregated and subsequently analyzed. It is essential to use a scale that subdivides a *chaîne opératoire* into its constituents in order to identify all the alternatives possible, and thereby develop models applicable to archaeological observations. In addition, this approach to representing activities in all their sequential specificity accords well with the standard archaeological practice of working at the most elementary scales in the study of technical chains.

Database: Activities and Goods

At Khambhat, the most elementary of technical operations correspond to different stages in the fabrication of different bead types. They can be described by creating a record of activities and goods (or items). Activities are the technical operations themselves; goods are the things produced and consumed. Together, they determine the specific relationship between linked operations required to make particular goods. In terms of goods, it is essential first to distinguish different stages of fabrication; then different forms of labor, each form having a corresponding skill; and finally, the energetic goods: electricity and vegetal combustion, tools, instruments, and equipment conceived of as consumable goods.

Each elementary operation or activity must be quantified in terms of production and consumption, and thus for each the task is to define a conventional unit of measure. These units are easy to define when there is a restricting element (for example, a man, a furnace, a polishing drum). In such cases, the most natural unit of measure is yearly output by a single individual, or by a furnace, or by a polishing drum. For less homogeneous operations, as in heat/color treatment (to turn a carnelian bead red), which takes place in furnaces unevenly filled with beads of various dimensions, the consumption and production of beads can be measured against a standard of 10,000 beads so treated.

As to goods, most of them are measured by a "natural" unit: raw nodules and fuel are measured by the ton; electricity in kilowatt hours; beads by their different stages of manufacture in units of 1,000. Labor is measured in terms of workers/day.

Index cards specifying activities and goods were used to record information regarding the different techniques utilized in the production of each type of bead (sensu technical chains), their place in the technosystem, and how they were quantified. The resulting data were used to evaluate both the foundation of the ethnoarchaeological model and the calculations made for each activity. Unfortunately, space is too limited to provide in full all recorded details. The following is a descriptive summary of the principal elementary operations.

- *Mining of primary raw material:* 200 kilometers south of Khambhat, people from the Bhils tribe mine agate and carnelian nodules from the terraces of the Narmada.
- *Sorting:* At the end of the day, miners deliver the nodules they have collected to depots operated by contractors. Nodules are weighed and sorted by size and quality (assessed on the basis of color and homogeneity).
- *Drying:* Once the nodules are trucked to Khambhat, they are dried for some months (September–March) on workshop roofs to allow the water in the porous matrix of the stone to evaporate.
- *Heat-treatment:* Each workshop contains an open brick furnace in which the chalcedony nodules are heated. Raw nodules have a hardness of 6 to 7 Mohs. The purpose of heating is to improve the stone's condition for knapping.
- *Knapping:* Once cooled, all nodules are knapped using indirect percussion by rebound.

- *Grinding:* Beads of high quality are ground on carborundum wheels; low-quality beads undergo initial grinding in wooden drums.
- *Drilling:* Beads of all qualities are perforated with the aid of a drill bow.
- *Polishing-shining:* Nonfaceted beads of all qualities are polished and shined in wooden drums. During polishing, beads are rotated in a mixture of water and carborundum dust for 24 hours. During shining, beads are rotated in a mixture of water and chalcedony powder (1 kilogram/drum) for 48 hours. Faceted beads of both low and high quality are polished and shined by hand on electrical wheels. Polishing wheels are made of lac and carborundum dust; shining wheels of lac and chalcedony dust.
- *Heat-color treatment:* Repeated heat-treating of carnelian beads (yellow chalcedony before heating) turns them red. This change in color is due to the oxidation of iron particles in the chalcedony. Carnelian beads are heated two to three times during different stages of production (generally, once after knapping and again after polishing and shining).

This ensemble of activities, or elementary operations, is organized as technical chains that vary according to the kinds of beads produced. In our study, we focused primarily on the main types of beads worked at Khambhat in early 1990. Bead types are defined on the basis of raw material, the quality of knapping, and the dimensions and morphology of the beads produced. Table 3.1 lists the bead types and their corresponding technical chains (the abbreviations used here are employed in all subsequent figures).

Analysis of Production Networks (or Technical Chains)

Technosystems often contain production networks—that is, technical chains or associated activities—that are relatively independent of one another, although there is competition for the consumption of primary resources. This is the case at Khambhat: networks for the production of different bead types compete for the labor of specialists, the use of furnaces, the consumption of energy, or the use of polishing drums. Nonetheless, these are autonomous networks. They are not linked through the circulation of goods, and therefore may be considered well-defined subsystems of the global system.[3]

Activity analysis should allow us to highlight the different networks (technical chains) in sequence in order to compare resources mobilized and financial profit (set by market prices). Such analysis is based on technical descriptions obtained by observing productive acts and by organizing and measuring each se-

Table 3.1.
List of Principal Bead Types and Their Corresponding Technical Chains

Bead type[a]	Mine	Sort	Dry	Heat	Knapp	Grind	Drill	Polish/ Shine	Color/ Heat
qI ls cal	x	x	No	12 h	x	Drum + wheel	1 drill	Drum	No
qI fct cal	x	x	No	12 h	x	Drum + wheel	1 drill	Wheel	No
qSp ls cal	x	x	Yes	24 h	x	Wheel	1 drill	Drum	No
qSm ls cal	x	x	Yes	24 h	x	Wheel	2 drills	Drum	No
qSl ls cal	x	x	Yes	24 h	x	Wheel	3 drills	Drum	No
qSp fct cal	x	x	Yes	24 h	x	Wheel	1 drill	Wheel	No
qSm fct cal	x	x	Yes	24 h	x	Wheel	2 drills	Wheel	No
qSl fct cal	x	x	Yes	24 h	x	Wheel	3 drills	Wheel	No
qI li rg	x	x	No	12 h	x	Drum + wheel	1 drill	Drum	Yes
qI fct rg	x	x	No	12 h	x	Drum + wheel	1 drill	Wheel	Yes
qSp fct rg	x	x	Yes	24 h	x	Wheel	1 drill	Wheel	Yes
qSp ls rg	x	x	Yes	24 h	x	Wheel	1 drill	Drum	Yes
qSm fct rg	x	x	Yes	24 h	x	Wheel	2 drills	Wheel	Yes
qSm ls rg	x	x	Yes	24 h	x	Wheel	2 drills	Drum	Yes
qSl fct rg	x	x	Yes	24 h	x	Wheel	3 drills	Wheel	Yes
qSl ls rg	x	x	Yes	24 h	x	Wheel	3 drills	Drum	Yes

a. qI, inferior quality (length 1–2 centimeters, 0.5–1 centimeter in diameter); qS, superior quality; p, small (length or diameter 1.5–3 centimeters); m, average (length 3–7 centimeters or diameter 3–5 centimeters); l, long (7–12 centimeters or diameter 5–7 centimeters); ls, nonfaceted; fct, faceteted; cal, chalcedony; rg, carnelian.

quence of production. Nothing at this stage is specific to Khambhat itself, save the manner in which work is done.

Production networks are made comparable by assuming that each mobilizes 1,000 workers for all tasks taken together (from extraction to the final stages of production). By normalizing different networks in this way, they may be compared in terms of the proportion of different operations (activities) involved, as measured by the number of individuals mobilized at each stage (extraction, knapping, grinding, drilling, polishing, and shining).

In figure 3.1, production networks are compared in terms of profit margin, calculated on the basis of one individual undertaking all productive tasks. What is striking here is that beads of superior quality, especially the largest beads, are significantly more profitable than those of inferior quality. The latter appear to be a weak source of profit though they are able to provide salaries for a significant number of workmen. The difference in profit margin between these types of beads is explained by their selling prices, which vary substantially from one type to another, and by expenses which, in comparison, vary little (figure 3.2). Selling prices depend on three factors: the nature of the raw material (carnelian or chal-

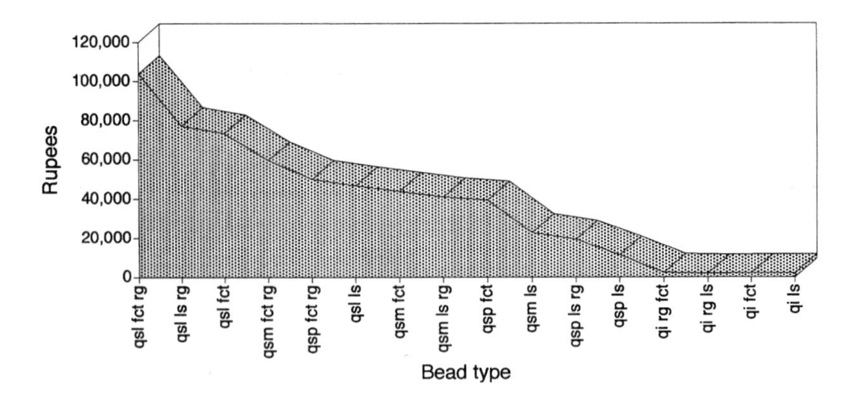

Figure 3.1. Annual profit margins per worker, by bead type (in rupees).

cedony), bead size, and the process by which they are polished and shined (that is, by hand or by drum). The main expenditure is salaries, and these are equivalent among different categories of craftsmen. Consequently, when one normalizes the selling price per individual, costs are held relatively constant.

Inferior-quality beads can be produced on a vast scale because of the technical procedures employed (specifically, use of drums for polishing operations). When measured in terms of 1,000 workers per year undertaking all stages of production, the production of inferior-quality beads is ten times higher than the production of beads of superior quality (figure 3.3). Technical procedures specific to the manufacture of low-quality beads are reflected in the proportion of workers employed in different tasks (figure 3.4): knapping is more rapid not only because of the small size of the beads, but also the technique used (particularly the use of drums during the earliest stage of grinding, as well as during subsequent polishing and shining).

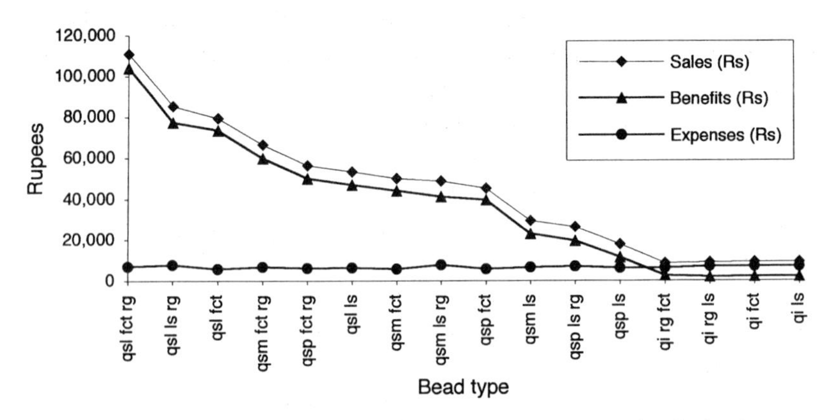

Figure 3.2. Annual sales, profits, and expenditures (in rupees) per worker, by bead type.

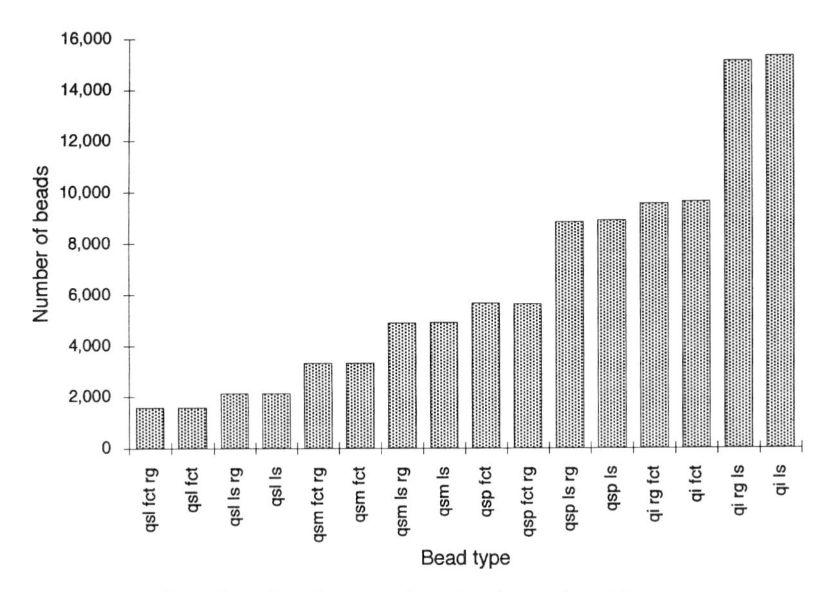

Figure 3.3. Annual number of beads per worker, all tasks combined, by type.

This last point, it would appear, is unique to modern conditions of production: the annual yield of beads per craftsman falls between 1,600 and 16,000 objects.[4] These high levels of productivity are explained by the use of electricity at certain stages of production and by the division of labor practiced. This level of productivity is not found in archaeological situations.

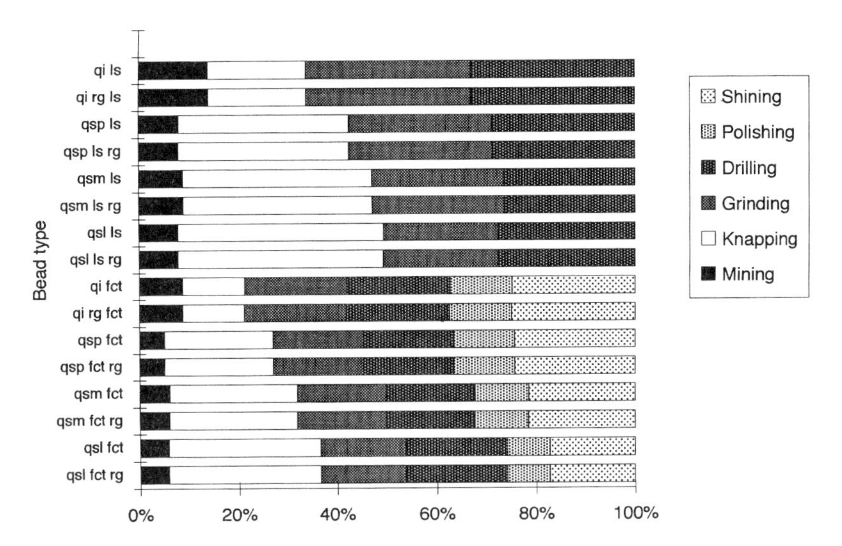

Figure 3.4. Percentage of workers at each task for the yearly production of each bead type manufactured by a total of 1,000 workers.

DISCUSSION

The technosystem of present-day Khambhat is not presented in its entirety since the elementary operations and the technical chains identified serve only as a referent for interpreting the archaeological situation. Briefly, a (bead) technosystem can be summarized in the following terms:

- *Elementary technical operations:* These operations integrate at the scale of a technical activity. They are the invariant factors upon which ancient technical chains can be constructed, regardless of cultural idiosyncracies.
- *Production networks or technical chains:* These correspond to the suite of associated activities followed to fabricate a unique type of object (beads). Note that the same bead can be made through several networks. As well, a network consisting of a specific set of associated activities allows for alternatives (for example, manual or mechanical polishing). Technical chains for bead manufacture are not specific to Khambhat. Rather, quantifying relevant production and consumption activities enables us to establish a point of reference upon which an ancient technosystem can be constructed.
- *The technosystem:* This global system corresponds to the complex articulation of elementary technical operations (that is, an association of production networks). The technosystem is described on the basis of technical chains and contextual data that give structure to the whole production network.

As the preceding discussion shows, findings from the ethnoarchaeological study are not limited to Khambhat's technosystem. Rather, they provide insights on invariant elementary technical operations relevant to our archaeological situation. This enables us to assess with precision the technical part of the overall system that holds constant over time.

ARCHAEOLOGICAL APPLICATION

We are now ready to look at the Harappan situation and to reconstruct it in the terms defined previously, that is, elementary technical operations, production networks, and technosystem. Such a reconstruction brings into relief the evidential gaps that make such analyses difficult. However, on the basis of what is known of the archaeological context, it is possible to reconstruct the elementary operations and networks, which, in turn, permits hypotheses about the global technosystem.

Elementary Operations

When used to interpret archaeological data, the ethnographic model is deceptive at the scale of elementary operations, because it presupposes the invariance of elementary operations, past and present. We pointed out at the beginning of the chapter, however, that it is possible to control this invariance through a comparison of archaeological, ethnographic, and experimental data.

As a result, we modify the model as follows to make it applicable to archaeological observations:

First, low-quality and faceted beads are not considered here. Harappan beads whose fabrication might be comparable to that of contemporary low-quality beads are the disc type. During the second half of the third millennium, these beads were of anecdotal importance compared with the bulk of beads produced. Faceted beads are so rare that they do not have to be considered here.

Second, to make a referent of nonmechanical operations it is necessary to consider techniques practiced in Khambhat prior to the introduction of electricity as well as certain mechanical operations still practiced in Yemen. In fact, operations in India and Yemen allow us to envision different grinding, polishing, and shining techniques that might have been practiced in the Indus valley.

In this study, three operations of internal coherence aggregate under the acronym "GPS" (grinding, polishing, shining). There are three categories of GPS, which correspond to the three primary combinations envisioned in Harappa:

- Manual GPS (*GPS-man*): This ensemble of operations employs grindstones and is quantified by ethnographic observations made in Yemen (see Inizan, Jazim, and Mermier 1992).
- GPS with bow (*GPS-bow*): This ensemble of operations involves the use of a polishing bow. This form of polishing was use in Khambhat before the arrival of electricity. The quantitative referent here is based on experiments undertaken by a Khambhat artisan still utilizing a polishing bow to work glass beads.
- Mixed GPS (*GPS-mix*): Our referent here is techniques still practiced in Khambhat before electricity (details in Trivedi 1964). These techniques were reserved for the production of small and midsized round beads. Grinding was accomplished using a grooved stone (showing functional use-wear traces) and polishing via a grooved wooden board. Shining was done by tumbling in a leather bag filled with emery and chalcedony powder.

Third, we must put aside the concept of "money" by restricting the model to purely material quantification (work, energy, and number of beads) and disregarding cultural values.

Analysis of Different Production Networks

In the observed ethnographic situation (Khambhat), each bead type is produced by its own network of uniquely associated activities. This situation is unusual because, in general, the same object can be produced through alternative technical chains. We have explored such alternatives, corresponding to different GPS processes, for the archaeological situation. A bead type associates thus with several networks, each of them corresponding to a particular category of GPS.

As before, each network is measured by a standard of 1,000 workers for each main bead type: *qsp* (small), *qsm* (medium), and *qsl* (large). With the exception of the GPS (grinding/polish/shine) process, different technical operations are treated as separate specialized tasks. Recall that we are obligated to separate these tasks in order to envision alternative ways work might have been organized.

Figure 3.5 presents three possible production networks for the fabrication of medium-sized beads. We note first that the combined task GPS is by far the most important. For example, in comparing the number of workers required to knapp, the ratio is approximately 25 for manual GPS (*GPS-man*), 12 for GPS with a bow (*GPS-bow*), and 2.5 for mixed GPS (*GPS-mix*). Thus, GPS-mix is the most profitable in that it produces large quantities of beads. Figure 3.6 shows that GPS-mix is 8 times more productive than manual GPS for fabricating *qsp* (small) beads, and 6 times more profitable for *qsm* (medium) beads. The difference in productivity between GPS-man and GPS-bow, regardless of bead size, is twice as much.

In our view, the most time-consuming and least productive GPS (*GPS-man*) was the one most frequently practiced during Harappan times. However, the pro-

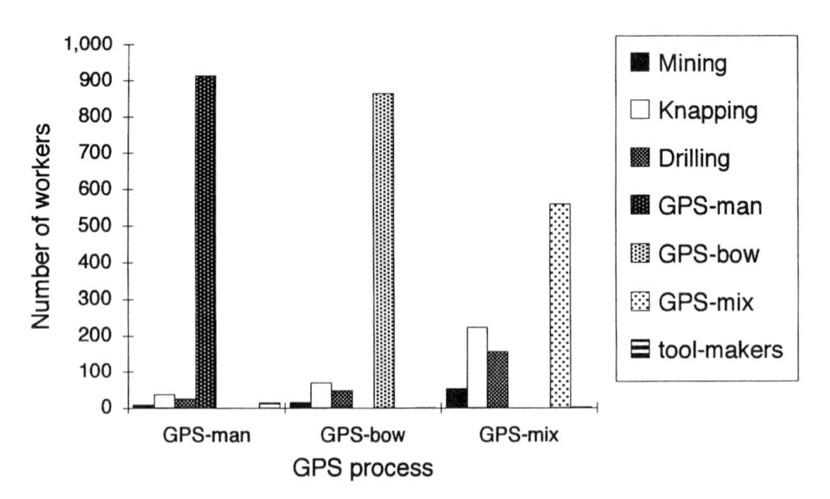

Figure 3.5. Comparison of number of workers, out of 1,000, dedicated to each task for the yearly production of *qsm* beads, according to the three GPS processes.

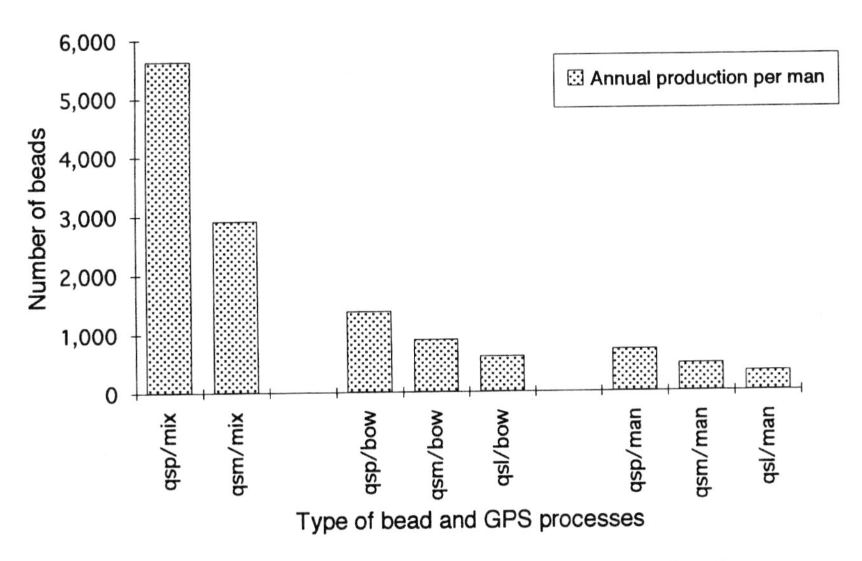

Figure 3.6. Annual number of beads produced per worker, all tasks combined.

ductivity of the GPS-man technique is not negligible when workers are employed full-time. For example, with all tasks combined (from extraction to shining), one worker can produce the following yields per year: by GPS-man = 723 *qsp*, 476 *qsm*, or 337 *qsl*; by GPS-bow = 1,383 *qsp*, 898 *qsm*, or 608 *qsl*; by GPS-mix = 5,634 *qsp* or 2911 *qsm*.

The Harappan Technosystem

In modeling the Harappan technosystem, it is best to make a detailed count of excavated archaeological materials in order to determine the absolute number of beads of different sizes (small, medium, large) and their different raw materials (carnelian, chalcedony). These specifics, understood as a constraint on production, allow us to estimate the number of workers required to fabricate a plausible number of beads, and following from that, to envision the socioeconomic situation in which this took place (regular versus occasional work; domestic versus workshop versus industrial context; independent versus attached specialists, etc.).

The difficulty here lies in the nature of the archaeological data. We do not know their representativity, either spatially (in terms of the excavated area compared with the site area or with an ensemble of sites) or temporally (in terms of the exact period of time during which beads were manufactured) (Gardin 1979).

To address this problem, we first attempt to estimate the total number of beads recovered from all Harappan sites; next, we estimate that percentage as it might correspond to the parent population, that is, in terms of the total number of beads worked during the Harappan epoch. This number can then serve as a basis for evaluating the hypothesized annual production of beads.

ARCHAEOLOGICAL OBSERVATIONS. It is difficult to ascertain the global number of stone beads recovered from Harappan sites. Many site reports made do with indications such as "a number of beads were found." However, if one trusts excavations at Mohenjodaro, Chanhudaro, Harappa, and Lothal—four sites yielding the largest numbers of recovered Harappan beads—the general sense is that the number of beads was rather limited (Kenoyer 1991; Mackay 1937, 1943; Marshall 1931; Rao 1979; Vats 1940; Vidale 1986, 1989).

Because it is difficult to arrive at an exact estimate of the extant number of Harappan beads, we must be satisfied with an overall estimate based on the number of large tubular beads, which, because they are exceptional, have always attracted the attention of excavators both in the Indus culture area and in Mesopotamia, where they are also found. It appears that the total number of Harappan beads does not exceed a few hundred, probably less than 500. Medium and small beads are always more numerous in proportion to large beads and are, *grosso modo,* 10 times and 100 times more numerous, respectively. On this basis we calculated the following estimates for extant Harappan beads: 500 *qsl,* 5,000 *qsm,* 50,000 *qsp.* The proportion of carnelian to chalcedony cannot be assessed without more detailed information.

REPRESENTATIVITY. How representative are recovered beads to the parent population, 1:100, or 1:1,000? We hypothesize 1:100 for the following reasons.

First, if excavated beads represent 1:1,000 in the parent population, that would imply that the original population was 500,000 *qsl,* 5 million *qsm,* and 50 million *qsp.* How, then, would we explain the loss of these millions of beads? It is a matter of material: on the one hand, beads are neither destructible nor recyclable; on the other, as pieces of jewelry, beads always attract the attention of excavators and pot hunters alike. This last point is important in the present. In today's antiquities markets, there are not millions of ancient beads for sale by collectors and antiquarians. This observation supports the idea that over the past centuries there has been a continuous market for them since their massive production in the third millennium. One could counter that, in fact, it is difficult to date carnelian beads whose shape has not changed over the millennia. However, there are beads whose shape or design leave no doubt as to their age and place of fabrication. In particular, there are diagnostic large tubular beads and "etched" carnelian beads, neither of which can be counted in the thousands.

Second, if the number of excavated sites compared with the number of known sites surveyed represents 1:1,000, the number of large sites excavated is proportionately quite high, on the order of 1 to 10.[5] Now, these large sites, though excavated extensively—Mohenjodaro was excavated on the order of 106,000 m^2! (Jansen 1994:267)—have never produced many beads.

Third, large tubular carnelian beads have been imitated in baked clay (for example, at Nausharo; Jarrige 1994:289), which allows us to suppose their production was limited for several reasons (perhaps it was a matter of restricted symbolic or physical access to large beads). Moreover, these large beads are often reused after breaking, which in a sense reduces the number of extant beads even further.

Fourth, intensive bead production results in a large amount of waste (chalcedony and carnelian flakes). From our observations and experiments at Khambhat, we determined for the Harappan case the following weights: 1 million *qsp* beads would have produced 11 tons of waste flakes; 1 million *qsm* would have produced 23 tons, and 1 million *qsl* 45 tons. If 50 million *qsp* beads had been produced, or even 5 million *qsm*, that would mean we should have recovered some 687 tons of waste from different Harappan sites. Although no actual figures on recovered waste are available, it appears that bead workers at large sites (for example, Lothal) never left more than a ton of waste (1 ton represents approximately 25,000 *qsl*, 50,000 *qsm*, or 100,000 *qsp*).

Thus, in adopting the hypothesis that excavated beads represent 1:100, we can conclude that overall Harappan bead production came to 50,000 *qsl*, 500,000 *qsm*, and 5 million *qsp* beads.

ANNUAL PRODUCTION. It is generally accepted that what is known as Harappan craft production lasted about 500 years (2500 to 2000 B.C.; Possehl 1993). Given this length of time, Harappan annual bead production would have averaged 100 *qsl*, 1,000 *qsm*, and 10,000 *qsp*.

Using these estimates, we can calculate for each polishing technique the annual number of specialized craftsmen involved in producing each category of bead (figures 3.7a, 3.7b, and 3.7c). This calculation does not take into account the number of workers mining nodules and those making bead-production tools, which in any case appears to be insignificant. In all cases, it appears there were only a few knappers and perforators: for *qsl*, the number of knappers is .02 and the number of perforators .01; for *qsm*, the number of knappers rises to .07 and .05, respectively; and for *qsp*, the figures are .4 and .3, respectively. For knappers and perforators, then, these counts signify that specialists devoted only a portion of their time to these endeavors, when compared with the known productive output of full-time specialists (6 hours/day, 26 days/month, 10 months/year). The number of abrader/polishers is clearly higher. For *qsp* beads, manual GPS requires 13 workers. When all tasks are considered together—that is, when we assume that all productive activities are undertaken by a single worker—the number of people implied in the annual production of Harappan stone beads using GPS-man rises to 16; using GPS-bow, it is 8.5 individuals; using GPS-mix, it is 2 (figure 3.8).

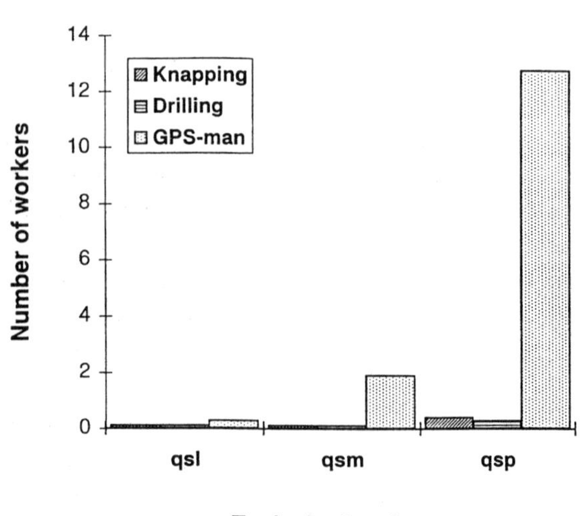

Figure 3.7a. Number of workers per task for the yearly production of 10,000 *qsp*, 1,000 *qsm*, and 100 *qsl* beads, employing GPS-man (by hand).

The fact that knapping and drilling techniques might have taken longer in prehistory than they do in Khambhat does not change these estimates. In fact (and as figures 3.7a, 3.7b, and 3.7c show), per year it takes less than one knapper and one perforator to produce 10,000 *qsp*, 1,000 *qsm*, or 100 *qsl* beads. Even if knapping and drilling techniques took 10 times longer, estimates of the total number of specialists implied in annual bead production does not alter significantly.

Our previous estimates allow us to propose two further calculations.

First, previous calculations suggest that if Harappan bead production lasted only 100 years rather than 500, then annual production would have been 5 times

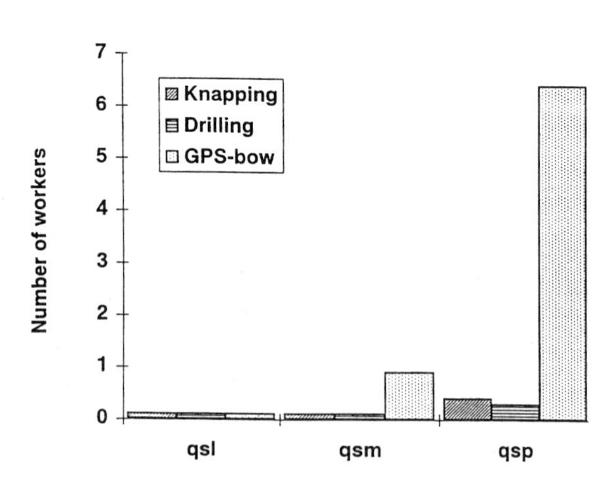

Figure 3.7b. Number of workers per task for the yearly production of 10,000 *qsp*, 1,000 *qsm*, and 100 *qsl* beads, employing GPS-bow.

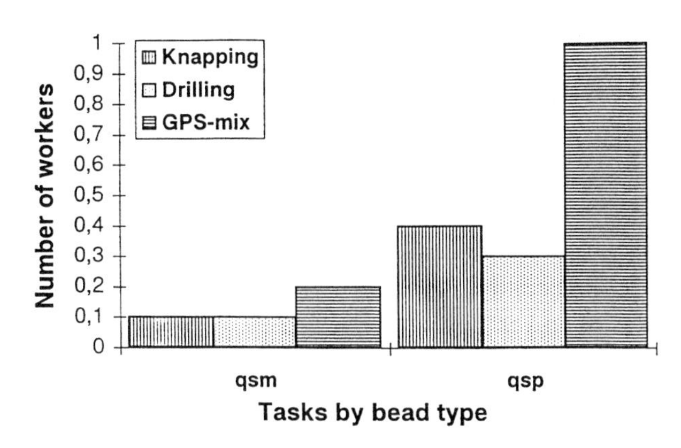

Figure 3.7c. Number of workers per task for the yearly production of 1,000 *qsm* and 10,000 *qsp* beads employing GPS-mix (grinding stone + leather bag).

more significant, yielding the production of 50,000 *qsp*, 5,000 *qsm*, and 500 *qsl* beads. This is remarkable: with manual polishing, the most time-consuming technique, the estimated number of artisans working full-time as bead specialists (all tasks combined) rises to 80. These 80 individuals would have been spread out among the several Harappan sites at which beads were manufactured. This again suggests the picture of an extremely limited number of specialists per site.

Second, there is archaeological evidence for a number of sites with traces of bead production (including Lothal, Mohenjodaro, Chanhudaro, Harappa, Dholavira, and Nagwada). These traces are mostly of small and medium-sized beads, except at Chanhudaro, where there is evidence for the production of large beads. Let us suppose that one bead specialist at Chanhudaro fabricated large tubular beads, and that, at some 20 other sites, two specialists fabricated all the small and medium-sized beads. This suggested division of labor by bead size is based not only on intersite evidence for differential bead production, but also on our study of technical skill. To the extent that each bead category requires a special set of

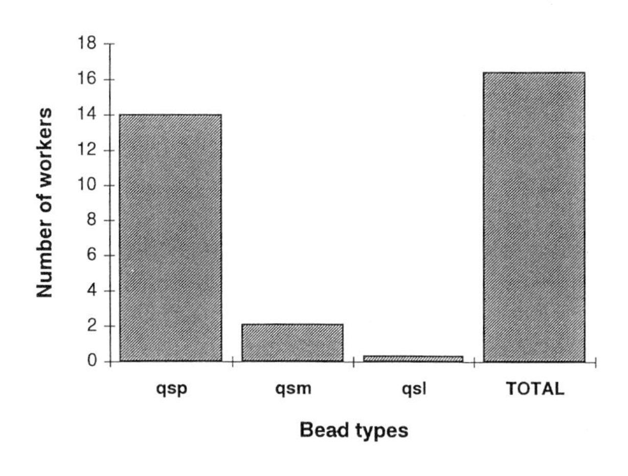

Figure 3.8. Number of workers, all tasks combined, involved in the yearly production of 100,000 *qsp*, 10,000 *qsm*, and 1,000 *qsl* beads, employing GPS-man (by hand).

technical skills that take time to learn, specialists adept at knapping *qsp* or *qsm* beads are not necessarily skilled at making *qsl* beads (Roux et al. 1995). If we also suppose that specialists using GPS-man worked full-time and that the bulk of bead production lasted only 100 years (four generations), the estimated results are as follows (for details, refer to figure 3.8): a single specialist at Chanhudaro undertaking all tasks combined could have fabricated 33,700 *qsl* beads (337 × 100 years); 40 specialists (2 specialists each at another 20 sites) working all tasks could have produced 2,169,000 *qsp* beads (723 × 30 workers × 100 years), and 476,000 *qsm* beads (476 × 10 workers × 100 years).

These calculations are somewhat less than the estimates proposed for overall Harappan bead production (50,000 *qsl,* 500,000 *qsm,* 5 million *qsp*). What is important, nonetheless, is that they indicate the possibility of a limited number of specialists producing a large quantity of beads in a limited amount of time.

Interpretation

This analysis of Harappan bead production has tested speculative hypotheses against empirical data. The following points obtain.

First, the number of specialists involved in producing stone beads was limited.

Second, specialists undertook the entire network, executing each of the individual tasks in a technical chain. If one considers the alternative hypothesis, of labor divided by task (as in Khambhat, where the various tasks are performed by different workers), then the estimates of labor drop to the point of absurdity (less than one individual, less than one perforator).

Third, in estimating the limited annual production of beads, specialists would have had other work to do, beyond bead manufacture. These other tasks would have been, in part, craft specialties, as one often finds other categories of objects associated with Harappan bead workshops: steatite beads and seals, shell artifacts, weights, and so forth. The common denominators, here, are the skills necessary for their manufacture, which are similar to those required for bead manufacture (that is, skills in knapping, drilling, and polishing; Pelegrin 1994). However, this polyvalence does not say anything about the frequency with which specialists devoted themselves to such tasks, and it is difficult to propose hypotheses as to their sources of income (in terms of direct or indirect access to agricultural resources).

Fourth, Harappan bead production appears so low that we cannot suppose a commercial enterprise in the proper sense of the term. We must now envision some kind of ad hoc productive system responding to a limited (religious, elite, or some other) demand. Production for elites allows us to suppose that stone bead specialists were "attached" sensu Brumfiel and Earle (1987).[6] Now we are in a position to question the theory of a "class" of specialists as currently hypothesized in the study of ancient urban societies and states. Was there a true "class" of specialists if it was composed of one or two workers per site? Is it reasonable to theorize that this "class" had a role in state formation? Indeed, the role of specialist is inferred in light of the question of how goods were redistributed and how redistribution was controlled. But where production appears to be ad hoc, there is no real problem with redistribution or the control of production and its appropriation. In fact, it is plausible to envision production developing as a response to the demands of established authorities.

Fifth, large tubular beads first appear during the second half of the third millennium. If knapping methods and associated techniques were those practiced during the preceding periods, these beads mark a technological innovation: the skills involved are new compared with the ones developed during the pre-Harappan period for the manufacture of small beads (Roux et al. 1995). These innovations are characterized by a long apprenticeship. What kind of demand could be strong enough, but nevertheless limited, to allow only one or two craftsmen to develop the skills necessary for the manufacture of these exceptional beads? In Mesopotamia, exceptional beads were dedicated either to kings or deities (André-Salvini 1995). Found in funerary or temple contexts, their function is known either through their excavated context or on the basis of ancient texts or engraved scripts. In contrast, in the Indus Valley large tubular beads have never been found in funerary contexts, but rather in settlements, workshops, or hoards. They are present at numerous sites, including small ones such as Allahdino or Chanhudaro. In addition, at most of these sites they have been imitated in baked clay. These data suggest that large tubular beads were intended for deities and not kings. On the one hand, this explains why these beads have not been recovered from funerary contexts but are found at most settlements. On the other hand, one can suppose that clay beads were intended to replace the originals in particular circumstances, such as when presenting the deities to the public (if one takes as an example some Hindu rituals). This could explain why clay beads are found in significant numbers, even on sites where carnelian ones were manufactured (such as Chanhudaro).

Finally, if large tubular beads were intended for Harappan deities, then they signal a hegemonic pantheon of deities over a large geographical area that transcended cultural (ritual) particularities. A hegemony of common deities would, then, characterize the Harappan civilization and could also explain a certain uniformity in material culture, likely also related to the dominant cultural system. The establishment of a new cult would have participated in the mechanisms underlying state formation with the new cultural hegemony stimulating the development of a new technoeconomic system. The formation of the state, then, could be the consequence of a cultural evolution, rather than founded on technoeconomic developments.

CONCLUSION

The aim of this study has been to reconstruct the technosystem of Harappan bead production on the basis of observations of an extant technosystem, Khambhat. It has allowed us to establish a descriptive procedure free of theoretical prejudice. The method employed, activity analysis, goes beyond the study of elementary technical operations and allows for a quantitative assessment of the whole technical chain. These calculations, in turn, make it possible to estimate the total number of craftsmen implicated in producing a certain number of objects. On the basis of these estimates, the ancient technosystem can be, in part, reconstructed.

The reconstruction of the technosystem of stone beads described in this chapter suggests, first, that from the moment craft becomes a specialized activity, it can produce a significant number of objects. Given this, the concept of "mass production" so often found in the archaeological literature dealing with the third millennium B.C. must be reconsidered. In fact, evidence of so-called mass production might reflect something closer to a specialized craft activity, one likely conducted at a low scale, rather than a "class" specialization. Second, concerning the formation of state-level societies, the development of a craft requiring specific technical skills appears to be the *consequence* of a preexisting sociological power rather than its cause. And third, the technoeconomic analysis of stone beads shows that it is possible to reconstruct their significance and intended purpose without assuming these are based on a priori social theory. Such a reconstruction enables us to identify the factors contributing to the formation of the Harappan civilization, which likely belonged to the cultural realm.

NOTES

This ethnoarchaeological study was supported by a grant from the French Institute of Pondicherry (Ministry of Foreign Affairs). Useful comments by the editors as well as by anonymous reviewers helped improve the earlier version. The chapter was initially translated by Marcia-Anne Dobres.

1. We wish to underscore here the importance of the French school of lithic technology, which has, for 20 years, endeavored to develop standard codes of analysis (based on experimental studies) permitting the interpretation of lithic industries in terms of techniques, methods, and skills. Combined with the concept of *chaîne opératoire,* these studies seek to reconstruct and understand technosystems and their evolution. A bibliography of this corpus of research is too large to provide here. Several especially representative works are CRA (1991); Inizan, Roche, and Tixier (1992); Pelegrin (1990, 1993); Perlès (1992); Pigeot (1990); Roux (1989, 1990). See further elaboration in Dobres (chapter 6, this volume).

2. The corpus of theories on urbanization and the development of the state are in almost unanimous agreement regarding the importance of specialized craft production. In particular, recall theses proposed by Johnson (1973), Tosi (1984), and Wright and Johnson (1975). For Tosi (1984), the concentration of craft workers in urban centers *necessitates* their control by an elite class and favors, as a consequence, the development of other classes and a bureaucracy. "For Johnson (1973), centralized administrative control of a specialized artisanal class and related phenomena creates a hierarchical class of administrators and [therefore] the State" (Francfort et al. 1989:23; authors' translation). For Wright and Johnson (1975), the production of craft goods, which become key objects of exchange, redistribution, and commerce, is one of the principal elements both driving and attesting to state-level organization.
3. The notion of networks is justified if the final production stage in a technosystem is conducted independently of others. In general, these are not found in agricultural systems, where production is well integrated.
4. The question of bead productivity has been analyzed as follows: we imagine a fictional workshop organized so that workers engage from beginning to end in all technical tasks necessary to fabricate beads. Then, the productive output of the workshop is determined per the total number of craftsmen. This allows us to determine the annual yield of beads per average craftsman in that workshop.
5. The largest Harappan sites, identified as such, number seven: Mohenjodaro, Harappa, Judeirjo-daro (Sind), Lurewala Ther and Ganweriwala Ther (Cholistan), Rakhigarhi (Haryana), and Kalibangan (Rajasthan).
6. "Attached specialists produce goods or provide services to a patron, typically either a social elite or a governing institution" (Brumfiel and Earle 1987:5).

REFERENCES CITED

André-Salvini, B.
1995 Les pierres précieuses dans les sources écrites. In *Les pierres précieuses de l'Orient ancien des Sumériens aux Sassanides,* edited by F. Tallon, pp. 71–88. Les Dossiers du Musée du Louvre. Editions de la Réunion des Musées Nationaux, Paris.

Brumfiel, E. M., and T. K. Earle
1987 *Specialization, Exchange and Complex Societies.* Cambridge University Press, Cambridge.

CRA (collective)
1991 *25 Ans d'études technologiques en préhistoire.* APDCA, Juan-les-Pins, XIème Rencontres Internationales d'Archéologie et d'Histoire d'Antibes.

Dantzig, G. B.
1963 *Linear Programming and Extensions.* Princeton University Press, Princeton, N.J.

Dobres, M-A., and C. R. Hoffman
1994 Social Agency and the Dynamics of Prehistoric Technology. *Journal of Archaeological Method and Theory* 1:211–258.

Francfort, H-P., M-S. Lagrange, and M. Renaud

1989 *Palamede: Application des systèmes experts à l'archéologie de civilisations urbaines protohistoriques.* Document de Travail No. 9. CNRS: LISH/UPR No. 315, Paris.

Gardin, J. C.

1979 *Une archéologie théorique.* Hachette, Paris.

Gwinett, A. J., and L. Gorelick

1981 Beadmaking in Iran in the Early Bronze Age—Derived by Scanning Electron Microscopy. *Expedition* 23(1):10–23.

Haudricourt, A. G.

1987 *La technologie science humaine. Recherches d'histoire et d'ethnologie des techniques.* Editions de la Maison des Sciences de l'Homme, Paris.

Inizan, M-L., M. Jazim, and F. Mermier

1992 L'artisanat de la cornaline au Yémen: Premières données. *Techniques et Culture* 20:155–174.

Inizan, M-L., H. Roche, and J. Tixier

1992 *Technology of Knapped Stone.* CNRS, Paris.

Jansen, M.

1994 Mohenjo-Daro, Type Site of the Earliest Urbanization Process in South Asia: Ten Years of Research at Mohenjo-Daro, Pakistan, and an Attempt at a Synopsis. In *South Asian Archaeology 1993,* edited by A. Parpola and P. Koskikallio, pp. 263–280. Suuomalainen Tiedeakatemia, Helsinki.

Jarrige, C.

1994 The Mature Indus Phase at Nausharo as Seen from a Block of Period III. In *South Asian Archaeology 1993,* edited by A. Parpola and P. Koskikallio, pp. 281–294. Suuomalainen Tiedeakatemia, Helsinki.

Johnson, G. A.

1973 *Local Exchange and Early State Development in Southwestern Iran.* Anthropological Papers No 51. University of Michigan Museum of Anthropology, Ann Arbor.

Kenoyer, J. M.

1986 The Indus Bead Industry: Contributions to Bead Technology. *Ornament* 10:18–21.

1991 Ornament Styles of the Indus Valley Tradition: Evidence from the Recent Excavations at Harappa, Pakistan. *Paléorient* 17:79–98.

Koopsman, T. C.

1951 *Activity Analysis of Production and Allocation.* Monograph No. 13 of the Cowles Foundation. John Wiley, New York.

Lemonnier, P.

1986 The Study of Material Culture Today: Toward an Anthropology of Technical Systems. *Journal of Anthropological Archaeology* 5:147–186.

1993 (editor). *Technological Choices: Transformation in Material Cultures since the Neolithic.* Routledge, London.

Leroi-Gourhan, A.

1943 *Evolution et techniques. L'Homme et la matière.* Albin Michel, Paris.

Mackay, E. J.

1937 Beadmaking in Ancient Sind. *Journal of the American Oriental Society* 57:1–15.

1943 *Chanhudaro Excavations 1935–36,* vol. 20. American Oriental Series, New Haven, Conn.

Marshall, J. H. (editor)

1931 *Mohenjo-daro and the Indus Civilization,* vols. 1–3. New Delhi.

Pelegrin, J.

1990 Prehistoric Lithic Technology: Some Aspects of Research. *Archaeological Review from Cambridge* 9:116–125.

1993 A Framework for Analysing Prehistoric Stone Tools Manufacture and a Tentative Application to Some Early Lithic Industries. In *The Use of Tools by Human and Non-human Primates,* edited by A. Berthelet, and J. Chavaillon, pp. 302–314. Oxford University Press, Oxford.

1994 Lithic Technology in Harappan Times. In *South Asian Archaeology 1993,* edited by A. Parpola, and P. Koskikallio, pp. 587–598. Suuomalainen Tiedeakatemia, Helsinki.

Perlès, C.

1992 In Search of Lithic Strategies: A Cognitive Approach to Prehistoric Chipped Stone Assemblage. In *Representations in Archaeology,* edited by J-C. Gardin, and C. S. Peebles, pp. 223–247. Indiana University Press, Indiana.

Pigeot, N.

1990 Technical and Social Actors: Flintknapping Specialists at Magdalenian Etiolles. *Archaeological Review from Cambridge* 9:126–141.

Piperno, M.

1973 Micro-drilling at Shar-i-Sokhta: The Making and Use of the Lithic Drill-Heads. In *South Asian Archaeology 1971,* edited by N. Hammond, pp. 119–129. Duckworth, London.

Possehl, G. L.

1993 The Date of Indus Urbanization: A Proposed Chronology for the Pre-urban and Urban Harappan Phases. In *South Asian Archaeology 1991,* edited by A. J. Gail, and G. J. R. Mevissen, pp. 231–249. Franz Steiner Verlag, Stuttgart.

Rao, S. R.

1979 *Lothal. A Harappan Port Town (1955–1962),* vol. 1. Memoirs of the Archaeological Survey of India 78. Government of India, New Delhi.

Roux, V.

1990 The Psychological Analysis of Technical Activities: A Contribution to the Study of Craft Specialization. *Archaeological Review from Cambridge* 9:142–153.

1999 (editor). *Cornalines de l'Inde: Des practiques techniques de Cambay aux techno-systèmes de l'Indus.* Editions de la Maison des Sciences de l'Homme, Paris.

Roux, V., B. Bril, and G. Dietrich
1995 Skills and Learning Difficulties Involved in Stone-Knapping: The Case of Stone-Bead Knapping in Khambhat, India. *World Archaeology* 27:63–87.

Roux, V., in collaboration with D. Corbetta
1989 *The Potter's Wheel: Craft Specialization and Technical Competence.* IBH, New Delhi.

Tosi, M.
1984 The Notion of Craft Specialization and Its Representation in the Archaeological Record of Turanian Basin. In *Marxist Perspectives in Archaeology,* edited by M. Spriggs, pp. 22–52. Cambridge University Press, Cambridge.

Tosi, M., and M. Piperno
1973 Lithic Technology behind the Ancient Lapis Lazuli Tade. *Expedition* 16:15–23.

Trivedi, R. K.
1964 *Selected Crafts of Gujarat: Agate Industry of Cambay,* vol. 5, pt. 7-A (1). Census of India 1961. Manager of Publication, Delhi.

Vats, M. S.
1940 *Excavations at Harappa,* vol. 2. Archaeological Survey of India, Calcutta.

Vidale, M.
1986 Some Aspects of Lapidary Craft at Mohenjodaro in Light of Recent Surface Record of the Moneer Southeast Area. In *Reports on Fieldwork Carried out at Mohenjodaro, Pakistan 1983–84. IsMEO-Aachen-University Mission,* edited by M. Jansen and G. Urban, pp. 113–149. Interim Reports, vol. 2. German Research Project Moenjodaro. RWTH and the IsMEO, Aachen, Rome.
1989 Specialized Producers and Urban Elites: On the Role of Craft Industries in Mature Harappan Urban Contexts. In *Old Problems and New Perspectives in the Archaeology of South Asia,* edited by J. M. Kenoyer, pp. 171–181. Archaeological Reports, vol. 2. University of Wisconsin, Madison.

von Neumann, J.
1945 A Model of General Economic Equilibrium. *Review of Economic Studies* 13.

Wright, H. T., and G. A. Johnson
1975 Population, Exchange and Early State Formation in Southwestern Iran. *American Anthropologist* 77:267–329.

4. Toward the Architecture of Household Service in New England, 1650–1850

ROY LARICK

VERNACULAR STUDIES IN HISTORICAL HOUSE ARCHITECTURE have shifted academic interest from large, overtly symbolic, high-status dwellings—those typical of Toynbee's "creative minorities" of history—to the modest houses and regional horizons of more common people. For New England, these studies have brought to light a wide range of moderate-sized houses and house types built during the late eighteenth and early nineteenth centuries. Vernacular architecture studies have also set the stage for understanding how, in the process of building moderate houses, individuals and families "actualized" various intentions within small-scale social and economic contexts.[1] As such contexts, early New England houses embody what Garrison (1991:197) terms the "working hypotheses by which people sought to manage their world and provide for the security of their loved ones." To modify Toynbee's concept, vernacular architecture studies have come to address the architecture of early New England's "creative majorities," and they provide a method for understanding how these majorities—the emerging American middle class—manifested social intentions in common dwellings.

This chapter is about a significant group of early New Englanders still left outside the actualizing paradigm: those who, at least for part of their gainful lives, lived below the threshold of house ownership and intentioned building. Until our own century, between 20 and 25 percent of this region's population was offered shelter and subordinate household membership in exchange for labor as "servant" or "help" (Fischer 1989; Izard 1996). If the complex social demography for this institution of "household service" pervaded rural New England houses for more than 200 years, the architecture of household service should present a fertile realm for social interpretation through technological analysis. To modify Toynbee's concept once again, it is time to extend historical analysis to the physical domain of New England's "subordinate minorities" of domestic architecture. This chapter shows how basic social arrangements of household service helped to configure the physical complex of rooms that sheltered a wide range of early New Englanders, and how house design in turn influenced social arrangements. By focusing on the technology for creating "service configurations"—the architectural space for household service—I hope to demystify the social acts of choosing among the technical alternatives by which early New Englanders built houses.

HOUSEHOLD SERVICE IN HISTORICAL CONTEXT

In little more than a decade, between A.D. 1629 and 1641, some 20,000 English Puritans settled the Massachusetts Bay Colony of eastern North America (Fischer 1989). Most had emigrated from East Anglia and contiguous counties to the south. So intense and historically significant was this translocation that it is called the Great Migration. Within a century, the descendants of the Great Migration had come to create a veritable "New England." Most immigrants arrived in households based on parent-offspring core units, with many having consanguinal and affinal extensions. These social units represented the economic and social middling classes of seventeenth-century Puritan England (for example, yeoman farmers, shopkeepers, clerical professionals), whose households routinely exchanged temporary subordinate members on a generalized reciprocal basis. The institution of household service therefore constitutes an important factor for social organization in early New England, and for the development of the region's unique housing tradition.

The Service Template in Seventeenth-Century England

The institution of household service provided a means to organize intensive household-based production in preindustrial agricultural and commercial endeavors (Wrightson 1982). In the seventeenth century, maintaining such a household as a viable unit of production required a great amount of labor, and many units periodically fell short. Young households could lack enough personnel even to make farming feasible. A husband and wife beginning domestic life together would typically wait 7 to 10 years for their first child to perform significant work, and several more years before the child showed competence in demanding tasks. Aging households suffered the loss of members to marriage, infirmity, and death. Recruiting outsiders into young and old households helped to stabilize their labor forces. Most servants came from "middle-age" households, units that had ceased to expand economically and thus might have offspring surplus to their reduced labor requirements. Functionally, the term "household service" connotes the range of tasks undertaken by servants primarily within agricultural contexts: cultivation and husbandry by men and boys; dairy and textile processing, as well as domestic maintenance, by women and girls.

The institution integrated diverse members of an agricultural society by numerous means. Taking an encompassing view, the Puritan service contract essentially involved two households: donor and recruiter. While relationships varied between these units, most held similar social and economic status (Laslett 1965). Consequently, the acts of giving and receiving subordinate household members maintained a generalized form of interfamilial reciprocity. Families participating in household service could ally themselves socially and economically, somewhat akin to exchanging offspring more permanently in marriage. From the recruiter's perspective, household service increased the size of the production unit at minor expense. From the donor family's perspective, household service was a means to socialize and enculturate offspring outside the home and to reduce domestic expenses.

At the institution's base lay a complex set of social interrelationships that mediated low-status servants with high-status masters. The potential for servitude was defined largely by age and marital status: able-bodied but not yet married. Servants were thus relatively youthful, most were between 15 and 24 years of age. Kussmaul (1981:3) estimates that youths regularly engaged in household service constituted about 60 percent of this age group. As youths, servants performed essentially adult tasks without full profit or ownership. Household service thus delayed the point at which many individuals married and accumulated property; older individuals and families could more easily consolidate material holdings. While servants indeed lay at the bottom of a complex social and economic hierarchy, the major determinant of their status, youthfulness, was short lived. Most servants could look forward to the day when they themselves would marry, own a dwelling, and then recruit their own subordinate householders.

Veering off the Template in New England

Household service evolved with the effects of capitalization and industrialization during the seventeenth, eighteenth, and early nineteenth centuries. In England, cash remuneration became important early in the seventeenth century, and it raised the economic threshold for recruiting servants (Kussmaul 1981). While wealthier households were able to hire more personnel, fewer total households could afford to take in servants. The recruiting process became more formal, anonymous, and much more differentiated by class. In general, the increasing monetary and market-driven aspects of the master-servant relationship—commoditization—helped to break the institution's generalized reciprocal base. In New England, nevertheless, the Great Migration seems to have set commoditization back for a time, in part by design. Wishing to encourage a self-reliant and self-contained society, Puritan leaders discouraged families from emigrating with more than one or two servants (Fischer 1989:28). Consequently, the institution's basis in generalized reciprocity remained an important fact of economic and social life well into the seventeenth century.

Household service in New England did veer from the English form. St. George (1986) argues that as New England society began to focus on material aspects of life, it fast became more "man-centered" and less "God-centered." Consequently, more transparent face-to-face transactions between ostensibly rational individuals replaced older linkages more embedded and communal in nature. Of interest here, the average age of servants began to rise. By the late seventeenth century, English males arriving as servants in New England tended to be contractually indentured adults beyond their teenage years (Galenson 1981). During the late seventeenth and early eighteenth centuries, the introduction of African slaves was the second and more significant New World change (Sweeney 1984). This enslaved population constituted a labor pool much different than that of local youths; slaves typically resided within just one or two households during their entire lifetimes. Indentured servitude and slavery thus tended to remove demographic youthfulness from the serving population and generalized reciprocity from the service contract.

Only recently have American historians begun to investigate the actualizing arrangements of household service in ways useful for technological analysis. Using popular journals of the period and the personal diaries of numerous young women, Izard (1993, 1996), Lasser (1982), McMurry (1984, 1988), Nylander (1993), and Ulrich (1982, 1990) have reconstructed female perspectives on domestic life in rural households during the late eighteenth and early nineteenth centuries in New England and New York. They all identify a significant dynamism in the service relationship. For example, as New England's economy grew, as market forces developed, and as population sectors became differentiated by wealth, "service became primarily a class specific, wage relationship, even while employer and employee still shared both home and workplace" (Lasser 1982:i). Through outright servitude, economically disenfranchised individuals could now seek room and board in addition to a small wage; wealthy households could demonstrate high social status by sheltering low-status individuals.

In the rural sector, the families that still could afford servants now attempted to hire from outside familial and community realms. While such anonymous laborers may have worked along with family members, neither side considered them integral household members. Farm and domestic work increasingly fell to gender-specific familial units (such as husband-son and wife-daughter), which had to get by with seasonally hired hands. Replacing the practice of household service and its underlying family alliances, the ideology of family self-sufficiency came to focus on the working roles of nuclear family members, especially housewives.

Nowhere did these roles change so quickly as in New Englanders' westward migration during the early nineteenth century. While this dispersal was neither religiously motivated nor as demographically or ethnically homogeneous as the Great Migration, many complete households left for new lands. With household service now in decline, few individuals called "servants" migrated with "masters," and even more tasks fell to frontier housewives. The new domestic ideology that only recently had come to segregate servants and their activities from the rest of the household now began to segregate physical work spaces as well as the social places of housewives and daughters from husbands and sons. These new social and economic arrangements helped bring on their own set of "post-service" architectural configurations.

HOUSEHOLD SERVICE AND HOUSE CONFIGURATION

From the actualizing perspective, early New England houses may be seen as a series of utilitarian components that had to be fit together in appropriate fashion. Functionally, a house held two physical sectors of greater and lesser formal organization (figure 4.1). The more formal sector, the "living quarters," included the parlor, the hall (later called the sitting or living room), the front entry space (earlier, the lobby or porch; later, the entry hall), and the sleeping chambers (later called bedrooms). Most service activities, nevertheless, played out within a less formal sector, herein termed "the service": the individual rooms and connecting passages as well as the overall structural unit associated with subordinated labor and residence. In many seventeenth-century New England houses, the hall mediated both sectors, serving for the preparation and consumption of food as well as for general social interaction. All kitchen activities drew materials from the service, which comprised, minimally, a pantry (for storing dry goods), a buttery (for aging and fermenting liquids), and a cellar (for cold storage). By the early eighteenth century, however, cooking hearths and storage routinely were to be found in the service, now the primary work room of the house (figure 4.2). While still retaining its mediating role, the hall came to have more restricted social functions. The living quarters' garret, or attic, also served in storing foodstuffs as well as in sheltering subordinate householders at night.

While each sector comprised just a set of rooms and features, the period principles for configuring rooms with each other and for articulating the two sectors were very complex. In fact, the process for actualizing the social and physical arrangements of household service stood on the base principles for configuring

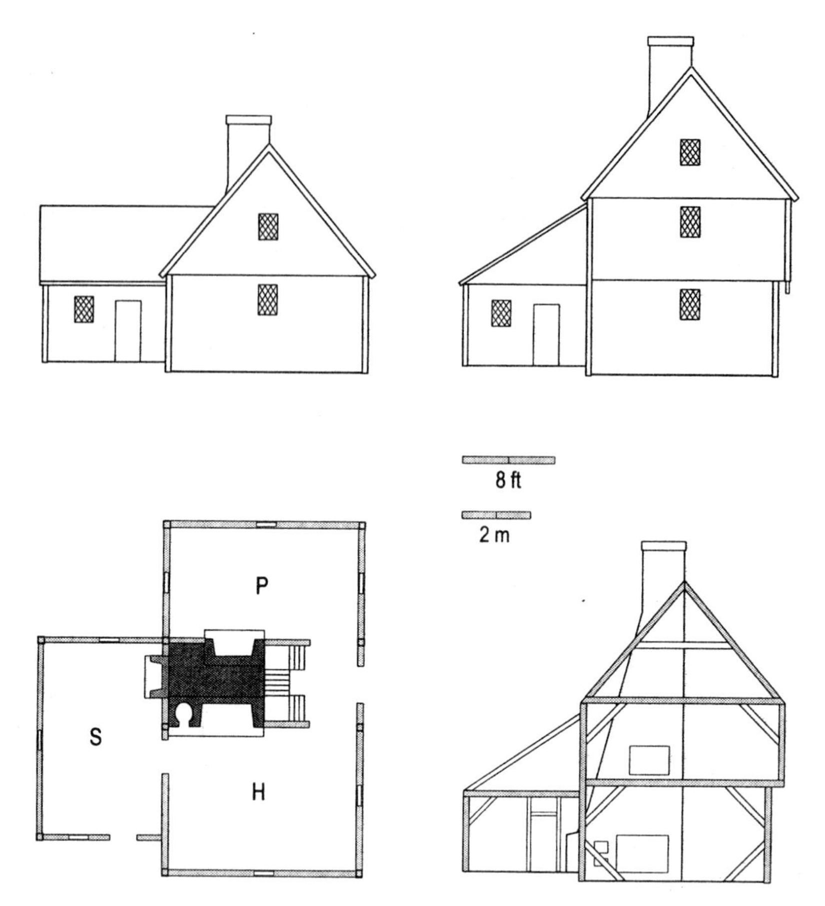

Figure 4.1. Side (gable) views and plan of elementary service configurations for mid- to late-seventeenth-century New England. P, parlor; H, hall; S, service. The one-story hall-parlor unit (top left) has a one-room service as an ell. The two-story unit has the same configuration, articulated in parallel, as a lean-to. The ground floor plan for both elevations (bottom left) shows cooking hearths (darkly shaded) in the hall and a small hearth (lightly shaded) later inserted behind the hall-parlor chimney. The lobby, with its staircase, lies between the chimney and the front door, within the chimney bay. The schematic frame truss (bottom right) shows simple bracing between posts and girts. The shaded frame members outline usable living space.

and articulating the architectural service. Ironically, base principles were rarely manifest in the conscious domain; they were not, for example, to be found in the manuals of period builders or in architectural pattern books. Base configuring principles seem to have remained unconscious or truly vernacular technological constructs. Not surprisingly, most historical architectural studies have utterly disregarded the structural units that sheltered the service: the outshuts, lean-tos, ells,

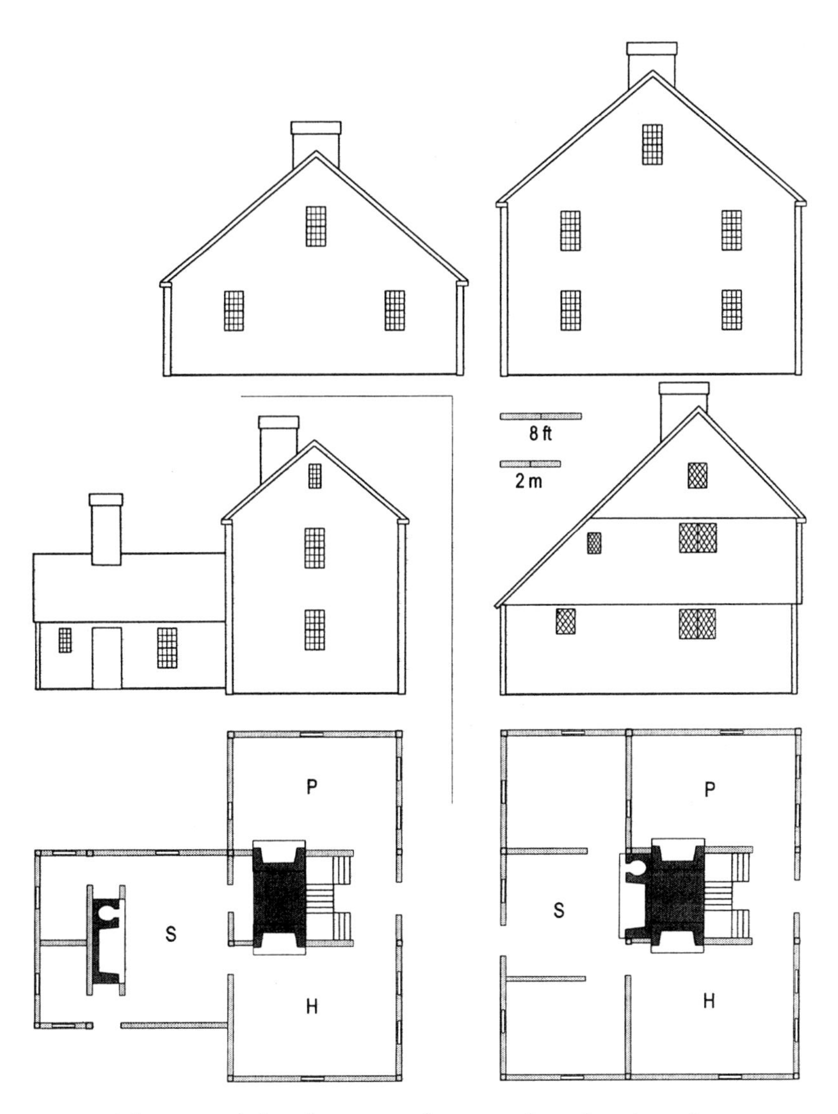

Figure 4.2. Side views and plan of service configurations for early-eighteenth-century New England. P, parlor; H, hall; S, service. The cooking hearths have been removed from the hall to the service. The two-story hall-parlor living quarters (left) has a one-story service ell with its own chimney. The full-length parallel-range plan (right) is shown in three elevations: early saltbox, "large," and Cape Cod (top left).

and wings commonly termed the "appendages" to house backsides.[2] In this analysis, I figuratively walk toward the rear of early rural New England houses to show that these components were integral to the physical structuring of the house as well as to the social arrangements of household service.

The Technology of Service Configurations

The goal of this architectural study of household service is to recover the period technology for configuring service rooms and for articulating the service and living sectors. This refers to an architectural or building technology with which the builders and owners of houses actualized socially appropriate service configurations. The formulation draws on Hughes' (1990) concept of "sociotechnical system" as a guide for articulating anthropological models of technology (Pfaffenberger 1992:493, 497). In early New England, it is these social aspects of technology that specified the base configuring principles for maintaining cohesive architectural structure and workable living arrangements. Some principles reconfigured the service around innovations in domestic chores. Others adapted new activities to the physical constraints of the architectural unit itself. Configuring principles also incorporated the possibilities and constraints of varied raw materials in the basic traditions and skills for building in wood, stone, and brick. Finally, configuring principles handled the problems of incorporating new ornamental styles onto old structural templates.

Reconstructing this technology can help us to understand how the builders and owners of houses handled continuity and change, how and why they adopted some new technical features and rejected others, and how they reviewed old structural principles for new uses. When observing the social arrangements for household service in early New England, one may see mortal change in the institution. However, when observing house configurations over the same period, one may see many slight adjustments in basic structural features that maintained a tensioned technological continuity. It appears that the architectural technology was more conservative—and thus more resilient—than the social institution it aided. The question is how builders and owners repeatedly drew from the pool of technical and symbolic alternatives to reintegrate household service into daily family life. The task in the remainder of the chapter, then, is to recover the basic principles that responded to actualizing issues with well-known technological alternatives.

A number of theories about historical contingency in the actualizing process help to put continuity and change into technological perspective. Regarding vernacular houses themselves, Glassie's (1975) structuralist analysis of Middle Virginia dwellings is the classic work. Glassie shows how basic architectural templates such as room volume and proportion remained nearly static for many decades while new ornamental styles frequently altered superficial appearances. In the concept of "technological style," Lechtman (1977) suggests that the technological principles behind basic structural templates are always largely unconscious and therefore slow to change. (Ironically, Wobst's [1977] contemporary conception of "style" placed it firmly in the dynamic—and self-conscious—actualizing mode, not in the realm of historical contingency.) Giddens' (1979, 1984) "structuration" specifies that antecedent conditions strongly influence the continuity and transformation of social action. In a related vein, Bourdieu's (1977, 1984) *habitus* identifies the socially constituted and historically based social structures that constrain the free will of individual agents. Ortner (1984) synthesizes these concepts into a "practice theory," which Dobres and Hoffman (1994) incorporate into the basic dialectic of structure and agency in technology.

In evolutionary theory, Gould and Vrba's concept of "exaptation" is appropriate to questions of technology. Gould and Vrba (1982; Gould 1991) recognize that some biological structures and behaviors are not related to current selection pressures. These features, having arisen by chance or prior contingency, nonetheless provide the phenotypic raw material for future adaptations. In other words, chance may produce physical structure and behavior that are in no particular adaptive relation to present selection pressures, and such features may be subsequently co-opted for uses removed from their original design. From the exaptive perspective, evolutionary process does not ordinarily create structures to fit changing requirements. Rather, biological and behavioral systems usually "make do" with preexisting resources that surface on their own.

Gould illustrates the general process of exaptation with an architectural metaphor from the Italian Renaissance: the "spandrels of San Marco" (Gould and Lewontin 1979; Gould 1993). He suggests that San Marco's spandrels (technically, "pendentives") are the architectural byproducts of mounting the cathedral's dome on top of four arches. "Spandrels" are thus not adaptations but chance occurrences co-opted to display evangelical imagery within a Christian cathedral.[3] The technological equivalent of exaptation may be seen in East Anglian configuration principles transferred to New England during the Great Migration. Here, preexisting technical alternatives seem to have been re-chosen and re-modified as the social arrangements of household service developed in time.

Macroeconomic Context

In conceiving a building technology for reconfiguring service arrangements in the face of social change, one must ask what was the larger origin and nature of those changes? All I have developed so far may be put in the period context of macroeconomic developments such as the capitalization of international trade, the nationwide industrialization of basic manufacturing, and the development of regional agricultural markets. Although these secular trends in some sense initiated or "drove" developments in household service and associated building technology, they had little to do with specific technological choices builders made. At the outset, nevertheless, one must recognize basic economic macrotrends in New England housing that provide the larger economic and social contexts for interpreting microlevel technological responses in service configurations.

Clearly, New England's economy expanded from the early seventeenth to the early nineteenth centuries. As a result, fewer houses were built small, more moderate-sized houses were built somewhat larger, and more were built with the compartmentalized plans previously associated with substantial houses. Consequently, the physical arrangements of household service became larger, more functionally specific, and more complex through time. A corollary factor here is the relative wealth of the building household. Smaller houses of any period accommodated service functions within fewer and more generalized rooms. Large houses tended to hold more rooms dedicated to specific uses. In focusing on moderate-size houses, I attempt to eliminate the social and technical uniqueness of extreme structure sizes while representing the largest reasonable sample of houses and household service arrangements.

EAST ANGLIAN SERVICE CONFIGURATIONS

Johnson (1989, 1993) analyzes changes in the structure and form of houses surviving from the fifteenth to the seventeenth centuries in East Anglia. He documents the varied arrangements among the three primary rooms of these dwellings (hall, parlor, and service) and explains the emergence of the composite three-room house of the early seventeenth century. This study makes clear that the evolution of the service and its articulation with the hall and parlor foreshadowed the service configurations of early New England. In the course of the following discussion, one must keep in mind a key concept in vernacular architecture—variety. There is no underlying ideal type of East Anglian post-medieval or New England premodern house. Owners and builders were always able to construct appropriate service configurations from often contradictory base principles in a number of ways.

A few basic tenets of timber framing underlie the technology of service configurations for all dwellings deriving from seventeenth-century East Anglia. Timber frames are a connected series of vertical sections called "trusses" ("bents" in New England) that give shape to and support the house. Conceptually, as a two-dimensional template, a truss has the shape of the house's gable profile (figures 4.1 and 4.3). Each truss comprises vertical posts and horizontal girts that further define interior floors and walls. As the basic framing template, trusses are erected along the length of the foundation, one at each gable end and usually two or three more spaced between. Horizontal beams are now used to connect the trusses, thereby completing the frame.

The space between two trusses is termed a "bay"; a house of five trusses has four bays. As its bays represent the usable volume of a timber-framed structure, the term "bay" provides a relative measure for the size and form of the structure as well as for the rooms within. Thus a four-bay house should hold more room(s) than a three-bay structure. In East Anglia, a long or cave side of the house always faced front with its entrance placed in a central bay. "Eave orientation" had its origin in the Roman Britain, and came to typify English houses through the Middle Ages and the Renaissance. It is only during the early nineteenth century that the American appeal to classical Mediterranean temples began to swing some New England houses to "gable orientation."

Closing Open Houses

In East Anglia of the fifteenth century, most middling households still built open houses of post-medieval open-hearth design, dwellings very different from any preserved in New England. Many held three rooms configured within a single four-bay structural unit (figure 4.3). The hall, the most important room, occupied the two middle bays and, at about the center of its floor, held the dwelling's single open hearth. The hall was open to the rafters with gable hip apertures serving to ventilate the open hearth. A system of braced vertical "crown posts" helped to strengthen the frame and provided visual ornament within the unceiled portion of the open hall (figure 4.3, top). In the "lower" hall bay (that adjoining the "service bay"), two entrances lay opposite each other and a light screen defined a transverse entry passage. Flanking the hall, the lower end bay held the service, a single room partitioned into work and storage areas. The "upper"-end bay contained the parlor. The service and the parlor typically were partitioned vertically, or "ceiled," to hold one or more upper floor "chambers" for sleeping and storage.

Studying the open house may yield insights into the nature of fifteenth-century service relationships and the structural principles for transforming service configurations during the sixteenth and later centuries. In being heated, in dominating physically, and in being centralized within the frame, the hall constituted the heart of house. Meals were both prepared and consumed there, and all householders generally interacted within. Alternatively, the parlor sheltered more intimate matters involving masters, but without heat this space had a limited function. Likewise, without heat the service saw only specialized activity regarding food and resource processing. Thus the spatial hierarchy of the full house recapitulated the social status of individuals and activities in their own domains. The parlor bay was framed longer than the service bay; the entry passage (subject to drafts) lay within the lower (shorter) hall bay; and the upper hall bay had larger windows. Nevertheless, the hall, in its entirety, served to mediate upper and lower ends of the house as well as dominant and subordinate householders.

During the sixteenth century, the open configuration was "closed." At the technological core of this transformation lay the chimney, a large coursed brick feature that contained one or two hearths. As a major new structural feature, the

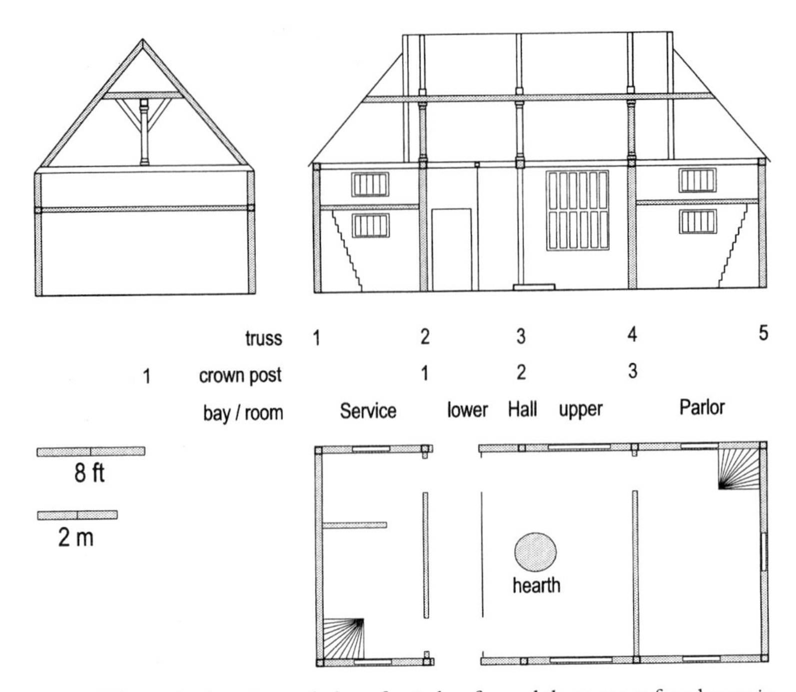

Figure 4.3. Schematic elevation and plan of a timber-framed three-room farmhouse in Brettenham, Suffolk, East Anglia (fifteenth century) (after Johnson 1989, figure 2). The hearth lies open in the middle of the hall; smoke vents at the hipped gables (top right). The shaded frame members outline the living spaces. Chambers lie above the service and the parlor, but not above the hall, which is open to the longitudinal spere beam (top). Gable view of frame (top left) shows the crown post bracing system. The staircases have radial treads (bottom).

chimney served to redefine the space and importance of the hall. It may be seen that a chimney could not be made to fit easily within a timber-framed structure. It had to be placed within a bay (between two trusses), and, since early chimneys were massive, one unit tended to take up much of an entire bay. The great range of surviving sixteenth-century chimney arrangements thus reflects the many variables and decisions caught in the act of closing the East Anglian post-medieval house (figure 4.4). Indeed, the variety of early closed configurations in East Anglia lies in contrast to the simpler designs found in the north and west of England, where masonry construction was more common. Chimneys in brick or stone are more easily and more compactly bonded to walls of brick or stone to produce more straightforward designs.

Adding chimneys and hearths effectively specialized the room functions seen in the open house, particularly those of the hall. While one chimney containing one hearth could be inserted into the hall in numerous ways (figure 4.4, top left),

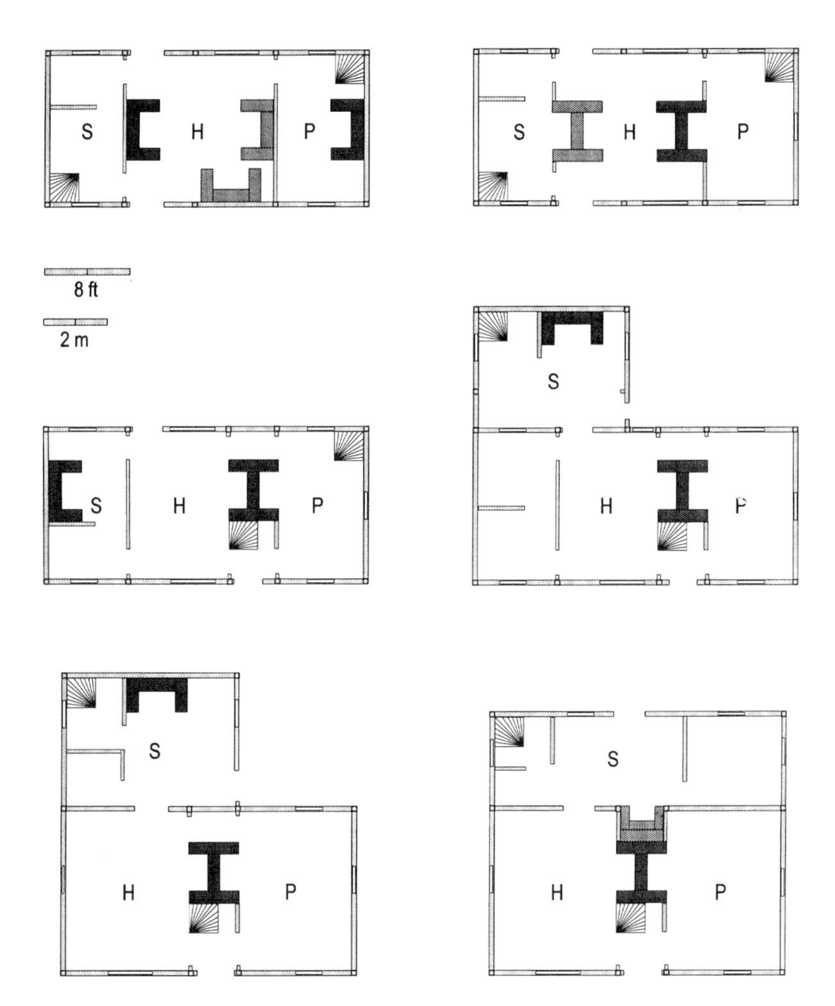

Figure 4.4. Alternative plans for inserting chimneys into timber-framed, three-room houses in sixteenth-century East Anglia (after Barley 1987; Brunskill 1970; and Johnson 1989, 1993). P, parlor; H, hall; S, service. The top plans show arrangements for "closing" a four-bay open house of the Brettenham type, with single-hearth (left) and double-hearth (right) chimneys. The middle plans show transitional types, in which the upper hall bay has been shortened (to become the chimney bay) to accommodate a double-hearth chimney, a lobby and staircase, and the front entrance. Bottom plans show the emergence of the three-bay hall-parlor, with perpendicular (left) and parallel (right) service articulation.

a chimney with back-to-back hearths could just as easily be placed within one hall bay to heat the two adjoining bays (upper hall and parlor, or lower hall and service) (figure 4.4, top right). Adding a second single-hearth chimney would allow all three rooms to be heated. If one was to take advantage of back-to-back hearths, the question became: which two rooms should be heated and which hall bay could be eliminated? The more common template placed a single chimney within the upper hall bay to heat the lower hall bay and the parlor (figure 4.4, middle left). In reducing the length of the hall to one bay, this configuration eliminated the need for complicated spere trussing. Moreover, in its one-bay, closed condition, the hall could now hold a chamber above.

Johnson relates the closing of the hall to developing hierarchical household relationships between masters and servants, parents and children, and husbands and wives. In his words (1989:199), "the lack of physical barriers in the open hall [acted] to obscure the inequalities of master/servant and gender relationships." Johnson (1993:109) thus sees the closing of the medieval hall as part of a shift from *Gemeinschaft* to *Geselleschaft* organization in early modern times. Heating the parlor and removing beds to chambers enlarged the private living space to which the husband and wife could retreat from the communal life of the hall. Adding kitchen facilities to the service eliminated one more function from the hall, further separating masters from servants as well as men from women (Johnson 1993:137–138).

Seventeenth-Century "Donor" Configurations

The early sixteenth century may thus be seen as a period in which carpenters experimented at closing houses by reformulating or actualizing new base principles in the real world. One of the problems facing period builders (and architectural historians) was the practice of reconfiguring the great variety of extant houses under incomplete or tentative principles. During this transitional period, then, carpenters had to close each standing house ad hoc. Thus, while structurally elegant, the four-bay plan of the Brettenham house illustrated in figure 4.3, for example, did not lend itself well to "closing." The plan's unitary longitudinal configuration could foster but limited communicative links among the three rooms.

Later and more successful configurations were "composite" ones, built from the ground up to configure the three rooms within two structural components. Composite configurations provided a socially more intimate and communicatively more efficient house. The composite plans were built on two sets of structural-functional linkages. On one hand, the chimney came to link the hall and parlor into a single unit of relatively high social status compared with the service. With the parlor now heated and the hall now ceiled, these two rooms took on nearly identical structures and similar functions. For example, in taking up much of the old upper hall bay (between the new hall and the parlor), the chimney effectively eliminated a room within. The remaining space came to serve as a centralized and more formal entrance area. Using this "lobby," one entered the house facing the chimney and turned left or right for one room or the other. A small radial-tread staircase could be set within the lobby (figure 4.4, middle). Finally, during the late sixteenth century, a new building template emerged to have two longer end bays (for the hall and parlor) flanking a shorter central "chimney bay" (for the lobby). This complete configuring principle but partial dwelling component may be termed the "hall-parlor unit" (figure 4.4, bottom).

On the other hand stood the close link between the service and the hall, which dictated that the service unit articulate to the rear hall side of the hall-parlor unit. This could be done in either of two ways. In one, builders attached the service perpendicularly to the hall-parlor unit behind the hall as an "ell," giving the entire dwelling the plan of an "L" (figure 4.4, middle right and bottom left; figure 4.1, left, for the early New England equivalent). Advantageously, a rear ell could be extended longitudinally one or two bays. In the other alternative, the service was articulated parallel to the hall-parlor unit as a rear structural range termed an "outshut" or "lean-to" (figure 4.4, bottom right; figure 4.1, right). Conveniently, a lean-to required minimal investment in construction. Both service configurations could be heated by inserting a third hearth into the rear of the hall-parlor chimney. In usurping kitchen functions from the hall (by receiving its own hearth and separate entrance), the late-sixteenth-century service finally gained functional and structural status nearly equal with that of the hall-parlor unit. By the early seventeenth century, then, moderate composite houses often had three heated ground floor rooms set within two structural units; ancillary service space was also to be had in the hall-parlor garret. These true composite configurations set the exaptive stage for early-seventeenth-century New England.

NEW ENGLAND TRANSFORMATIONS

With Johnson's East Anglian sequence for reconfiguring service and living quarters on the eve of the Great Migration, New England service configurations may begin to make sense. Kniffen's (1986) law of contemporary dominance suggests that only the newer early-seventeenth-century East Anglian configurations should have provided the basis for architectural planning and practice in New England. Without question, the two-unit, three-room, center-chimney configuration was the model, or template, for Puritan settlement (figure 4.4, bottom), but reality did not always accommodate the model. Thus it may be seen that the exigencies of migration forced many families to build small, abbreviated dwellings during the first decades of New England settlement. Although many of these early dwellings had but one or two rooms, they were still built on the base principles for two units and three rooms—many were later completed to fulfill their templates.

Most full-template houses had one or more unheated rooms on the ground and upper floors that linked closely with service activities (figure 4.1). For example, upper-floor chambers functioned for storage as much as for sleeping. Householders slept primarily on the ground floor: parents in the parlor; and children and servants within the hall, the kitchen, or the unheated rooms. Children and servants might use the upper-floor chambers when the ground floor rooms were too crowded, especially during temperate seasons. St. George (1986:349) suggests that New England's "up and back" service configurations reflected seventeenth-century symbolic household priorities. Domestic maintenance activities were to be consciously hidden from public view, behind and above the main living unit. However, a concern to remain out of public view may have been just one configuring principle; an equally critical dynamic hinged on the social arrangements of work. Having masters and servants share close quarters and undifferentiated spaces helped maintain the close-knit service relationship of the rounding period. The technological historical contingencies of articulating the service rearward of the hall-parlor unit may have equaled any social considerations. In other words, because seventeenth- and early-eighteenth-century New England houses expressed minimal spatial differentiation, the social arrangements of household service needed to remain intimate and flexible.

As the service contract became subject to commoditization during the eighteenth century, service relationships increasingly segregated servants from owners. Wealthy New Englanders were able to reconfigure their houses to suit the changing contract by differentiating them into functionally and socially specific sectors. In this period, the informal, organic way that post-medieval houses had been enlarged no longer made aesthetic or social sense. Clearly, the early compact configurations were not suited to the evolved forms of household service in which the daily lives of masters and servants became ever more segregated. Now, neoclassical concepts such as the Golden Section were used to create larger, more specialized floor plans and facades. For example, a hall-parlor unit could be literally doubled, front to back, to yield four heated ground floor rooms under a single roof and within a symmetrical package (figure 4.5, right). This was commonly termed "double-pile" construction.

Inside, the hall and the parlor were often set off from general circulation by replacing the lobby with an "entry hall" running the full depth of the hall-parlor unit (figure 4.5, bottom). The entry hall provided for circulation between the front door and the service without passing through the front rooms. Using the continuation of the entry hall on the upper floor of double-pile houses, servants could gain relatively private access to garret sleeping accommodations. Finally, a heated room was often built for dining. When configured within an ell articulated to a simple hall-parlor or "single-pile" living quarters, the dining room was placed behind the old hall and in front of the kitchen (figure 4.5, bottom left). When configured within a double-pile structure, the dining room was usually placed in the rear range behind the parlor and beside the kitchen (figure 4.5, bottom right).

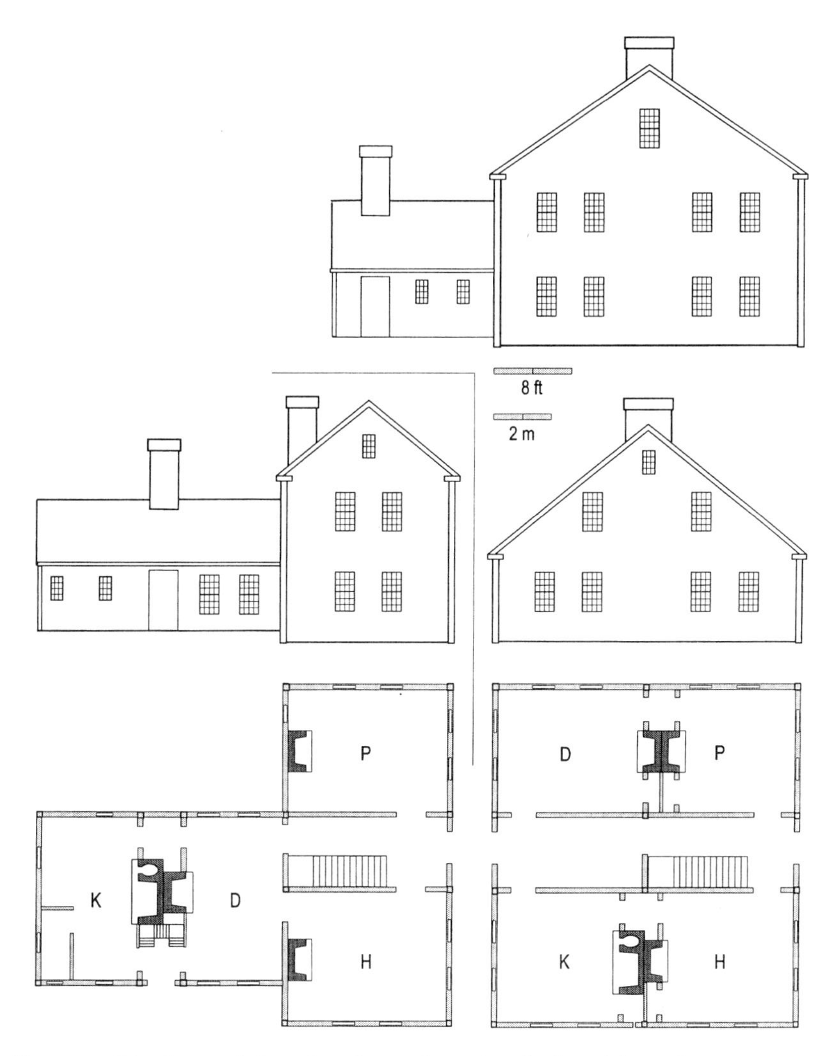

Figure 4.5. Side views and plan of service configurations for late-eighteenth-century "Georgian" New England. P, parlor; H, hall; D, dining room, K, kitchen. In both single-pile (left) and double-pile (right) versions, the old chimney bay now holds the entry hall. The single-pile upright has a one-story service ell configured as a small hall-parlor (three-bay, two-room) structural unit. The two-story double-pile structure (top right) has an old one-room half-house added as an ell to remove the kitchen from the main dwelling unit. The double-pile versions have significant subordinate sleeping space in their garrets. The two-room ell (left) has space in its own garret.

Configuring the entry hall between the parlor and the old hall helped to privatize both rooms. Configuring the dining room to the old hall also helped to specialize this room away from food and toward more restrained social interaction. Inserting both the entry hall and the dining room served to recapitulate the closing of the old hall to general circulation in sixteenth-century East Anglia. Both the sixteenth- and the eighteenth-century transformations came to make the old medieval hall, the single most important room of the premodern house, into a smaller more intimate space from which servants could be excluded. Throughout the centuries then, the size and social significance of the old hall was sacrificed to re-negotiate the dwelling and working spaces amid changing family and servant places.

As substantial houses received larger service sectors, the ell provided the best means to segregate work and to subordinate householders. In this regard, even large double-pile houses received ells by the end of the eighteenth century (figure 4.5, top right). By this date then, substantial houses almost always had composite configurations and a uniform ground floor plan. The front "upright" held two or four heated rooms, sometimes including the dining room. The rear ell held the kitchen and associated storage and work rooms, and often the dining room. The ell's upper floor, or garret, now could contain space for servant residence.

For the more moderate houses of the eighteenth century, change was less revolutionary. While facades were regularized and ornamental moldings were made prominent outside and within, the old center-chimney hall-parlor units articulated with small service units (parallel or perpendicular) remained the most common structural form of moderate dwelling. These houses persisted to shelter families that could not recruit or could no longer regularly recruit servants, their architectural conservatism reflecting an emerging but not yet dynamic middle class.

THE ARCHITECTURE OF HOUSEHOLD SERVICE IN COLLAPSE

As noted earlier, two economic macrotrends greatly changed New England's household-based means of production after the Revolution. Agriculture became more market-driven and western lands opened for rapid settlement. Along with the market forces, improved roads, canals, and then railroads essentially put an end to the cultural and geographic patterns that had previously guided colonial life. Even farm families who remained in the old colonies were eventually forced to compete with frontier settlers for an all-important share of the interregional agricultural markets. But competition often brought decent cash profits. In the wake of these trends, the institution of household service began to collapse. Wealthy families now hired help for wages, but the emerging middle class generally looked toward self-sufficiency in household labor requirements. In the post-service era, most households had little to do with live-in labor. Under these conditions, middle-class New Englanders spread a new type of moderate-sized post-service house throughout the northern tier of the country. Ironically, post-service houses show the greatest structural differentiation between living and working quarters.

Built on a slightly smaller scale than their Georgian predecessors and ornamented with newer neoclassical imagery, post-service houses frequently re-employed yet older configuring templates to achieve composite form. Thus two new upright plans came into use, each based on older exapted principles. First, a builder could eliminate an end bay from a three-bay double-pile structure to create a moderate-sized "half-house" upright (figure 4.6, top). The two formal rooms were now arranged front-to-back, with the entry hall traversing the upright along one side. From the front, the half-house upright exhibited an abbreviated bilateral symmetry entirely appropriate to period neoclassical fashion. (Figure 4.5, top, shows the front profile of a small one-story half-house.) While the plan was an old one (seen in sixteenth-century Europe as well as in eighteenth-century North American seaboard cities), the half-house upright became popular inland only during the early nineteenth century. Second, a compact form of the early colonial, parallel-range, three-room house provided the "temple" or "classic" upright (figure 4.6, middle). Similar to the half-house unit, key features of the temple upright included a moderate-sized structural package holding up-to-date room configurations. More important, the temple upright added fashionable if iconically ersatz classical imagery to the early-nineteenth-century house. In both eave and gable orientations and in one- and two-story elevations, a great range of middle-class families built houses having a temple upright.

Rearward, the post-service ell contained a heated kitchen and associated storage in addition to another heated room, most often for dining (figure 4.6, middle). Not many post-service ells had attic space for sleeping. For substantial dwellings in which servants still played important roles, the service floor area and number of rooms sometimes approached that of the upright itself. Full two-story ells could accommodate a heated dining room and kitchen as well as a number of specialized work and storage rooms on the ground floor (figure 4.6, top). Upper-floor space could be given over to full-height bedrooms as larger households might host live-in "hands," paying borders, and occasional boarding in-house wage laborers. In the end, however, most large and moderate houses were differentiated more around the relative size of service space than in its organization.

The changing domestic activities of middle-class men and women also helped to reconfigure moderate houses during the early nineteenth century. At the level of the individual farm, such developments brought specialization and an ever increasing focus on cash crops and home processing (Hubka 1984, 1986). McMurry (1984, 1988) documents how, under these conditions, farm women took

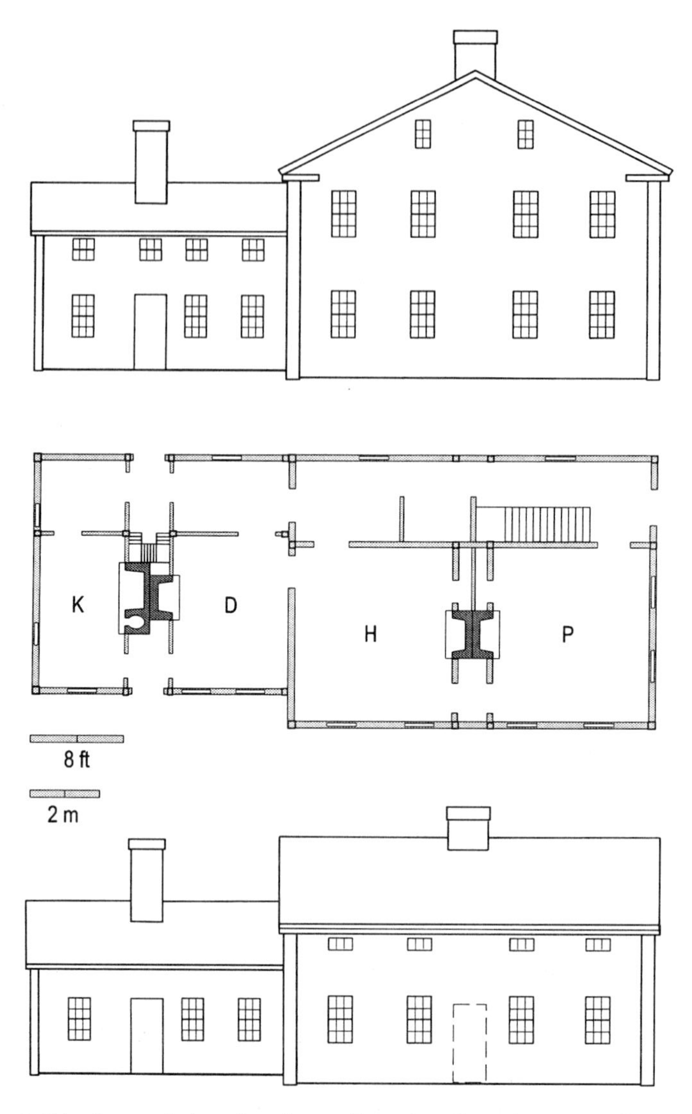

Figure 4.6. Side views and plan of service configurations for early-nineteenth-century New England. P, parlor; H, hall; D, dining room; K, kitchen. Both uprights have two formal rooms arranged front-back (right-to-left on the plan), connected to the service by an entry hall alongside. The two-story half-house upright (top) is framed as a truncated (two-bay) double-pile structure. The story-and-half template (bottom) has a gable-oriented, three-bay, parallel-range plan. The two ells, while varying in volume, have similar three-bay, parallel-range plans. In configuration, they are small, seventeenth-century-type structures with updated framing and finish. The bottom house plan was also popular in eave orientation, with the upright front door centered in front of the chimney on an eave facade (broken line).

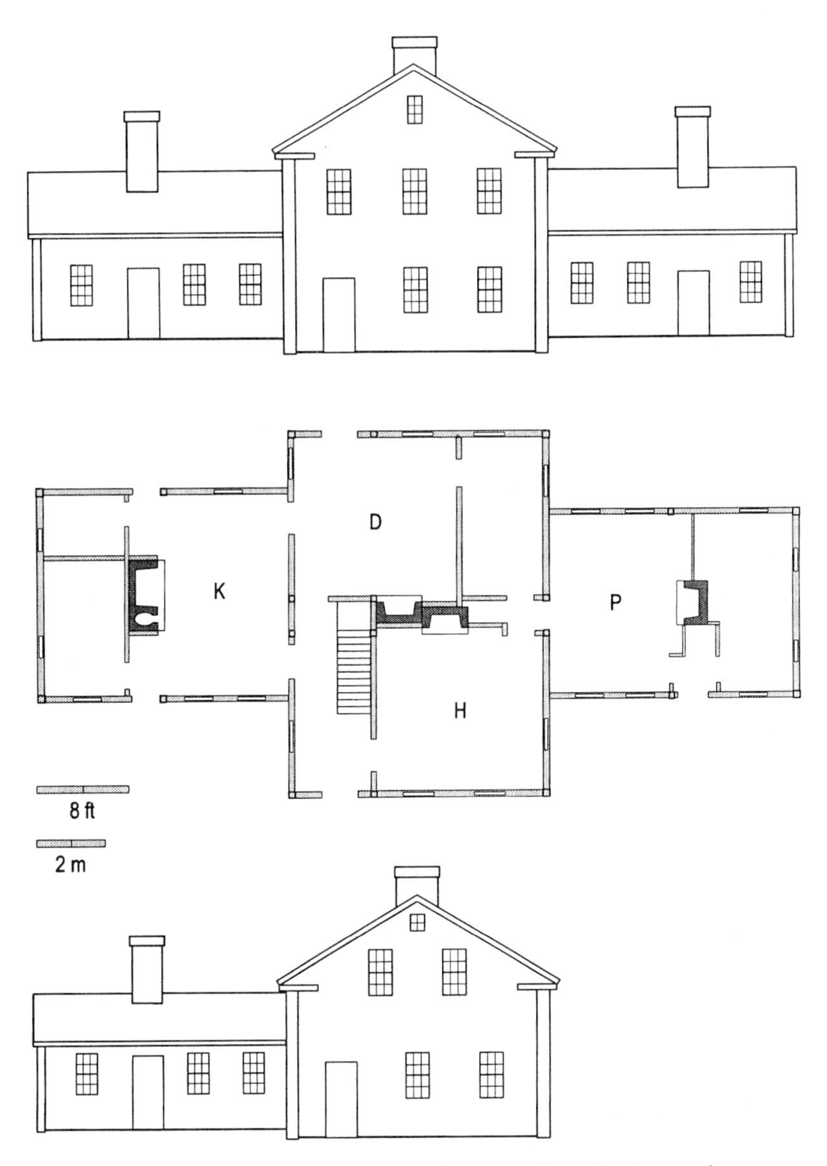

Figure 4.7. Front views and plan of service configurations for early-nineteenth-century "post-service" New England and Western settlements. P, parlor; H, hall; D, dining room; K, kitchen. Both temple uprights have similar three-bay, parallel-range plans. The wings have two-bay, two-room plans with a heated room inside the chimney. By this time, service structures recall the most abbreviated house configurations of the early seventeenth century. For the very moderate-sized house (bottom), the upright front room is the parlor and the rear room doubles as hall and dining room. This house plan could also have eave orientation, making its wing project rearward as an ell.

active architectural roles in transforming dwellings to meet their own interests. It is in these contexts that the service came to form an integral component of the front facade as a lateral "wing" (figure 4.7). Ironically, amid the early-nineteenth-century fashion for austere classical Greek temple iconography, service wings were articulated according to lingering flamboyant Renaissance principles. In the sixteenth century, Italian architect Andrea Palladio had developed palatial dwellings in which wings housing formal living quarters bilaterally flanked a central pavilion. English colonialists subsequently revived Palladian designs for very substantial houses and civic buildings during the eighteenth century. Finally, with his own Monticello, Thomas Jefferson helped repopularize neo-Palladian plans across the South during the first decade of the nineteenth century.

In the North, wealthy owners adapted the neo-Palladian plan by flanking a large temple upright with wings (figure 4.7, top). One wing usually held the most intimate spaces of the living quarters; the other comprised the service. Alternatively, many owners of moderate means were more pragmatic with their display of Palladian concepts, expressing the essential principles within the confines of traditional templates. In flanking a moderate-sized temple upright with just a service wing, Palladian symmetry could be abbreviated to produce a simpler composite dwelling, one in which the working quarters became a facade component nearly equal with the upright (figure 4.7, middle). The composite structure and pleasing asymmetrical appearance foreshadowed the irregularly massed Victorian houses to be built later in the century. The great majority of these wings contained a kitchen, storage, and dining area. A front facing door and decorated doorway evinced the major role of this unit in the domain of the nuclear family.

DISCUSSION

In the introduction to *Servants in Husbandry in Early Modern England,* Kussmaul (1981:3) remarks that "most youths in early modern England (1574–1821) were servants; that so few are now is one of the simplest differences between our world and theirs." A nearly identical statement could be made for the corresponding period in New England. From the Great Migration through the mid nineteenth century, a wide range of families at times gave out subordinate workers and at times received others into their own households. The social hierarchies of service relationships have thus typified household life in this region for more years than they have not. For us, household service is best understood as a social institution responding to the needs of home-based productive units in preindustrial Europe and North America. Evidently, the arrangements of household service, asymmetrical in all realms, were always surrounded by tension. Consequently, the architecture of household service was continually subject to modification.

During the seventeenth and early eighteenth centuries, the service still provided a technological means to integrate serving and familial householders into one corporate group. By the late eighteenth century, nevertheless, wealthier families had readapted older configuring principles or templates to put space between themselves and their servants. As relations between recruiting and serving householders continued toward asymmetry, the rules of residence and activity became more complex and explicit, and service configurations came to manifest truly rule-bound architectural spaces. Ironically, as recruiters succeeded in diminishing the status of servants, they had to invest more in the architectural support of servitude. With the breakdown of household service in the mid nineteenth century, these formalized spatial arrangements came to structure the work routines of nuclear householders by age and gender alone.

The role of historical contingency in the actualizing process guides my interest in architectural technologies. As I have shown in non-Western ethnoarchaeological contexts, the innovative aspects of technology that both create and reaffirm world views often look to the past as much as to the future (Larick 1985, 1986, 1991). In applying older principles to a new house, New England owners and builders selected or co-opted structural solutions preexisting within their technological repertories. Recurring principles, such as parallel and perpendicular service articulations and structural abbreviations, provided the conceptual raw material from which builders and owners created new components such as service wings or temple and half-house uprights. As such, a cultural sort of evolutionary exaptation characterized this architectural technology. In exapting older structural forms and architectural features, house-building technology remained conservative and predictable and thus put little strain on the trade system in which boys learned the skills of building and in which men actually built houses.

The effects of migration also introduced historical contingency to the transformation of service configurations. While numerous alternatives for service configurations were known in early-seventeenth-century East Anglia, only the rear parallel and perpendicular service articulations survived in the New World. Ironically, while the parallel service articulations that typified early New England (especially in the "saltbox" and "Cape") derived from early-seventeenth-century East Anglia, these particular forms had never been popular there. Thus the opportunities and dilemmas of migration were met with both the elementary recapitulation and the creative re-use of older templates as well as with outright technical novelty.

In the final analysis, the technological system for building and occupying early New England houses remained dynamic even as its products could often appear conservative. Subdued architectural responses to the tensioned issues of service relationships were part and parcel of the dynamic changing nature of social and economic factors, including a shift from agriculture to an increasing industrial mode of production and the changing nature of servitude. My purpose in analyzing the technology or sociotechnical system for creating service configurations was to establish a robust theoretical underpinning of social agency and a historical contingency to the study of domestic architecture. Social acts with material consequences must always be interpreted between the opposing realms of social agency and historical contingency. Identifying the basic configuring principles for early New England houses provides an empirical context for interpreting the resilient social hierarchies that have pervaded American life for centuries.

NOTES

Old Sturbridge Village (Sturbridge, Massachusetts) granted the Research Fellowship under which this project has seen light. The Ethnoarchaeology Seminar (1994), Department of Archaeology, University of Calgary (Calgary, Alberta), generously provided a forum in which many ideas were refined. Historic Deerfield and its library (Deerfield, Massachusetts) have made many special resources available. Materials and advice have come through the Fairfield Historical Society (Fairfield, Connecticut), the Western Reserve Historical Society (Cleveland, Ohio), and the Woodstock Historical Society (Woodstock, Vermont). For their efforts in reading and discussing drafts for this chapter, I thank Abbott Lowell Cummings, Marcia-Anne Dobres, Chris Hoffman, J. Edward Hood, Matthew Johnson, Jack Larkin, Bob Paynter, and Nancy White.

1. Vernacular architecture studies have a long history in Britain. Among recent works, Brunskill's handbook (1976) and Johnson's East Anglian studies (1989, 1993) relate directly to the present analysis. Glassie (1968, 1975) and Kniffin (1986) pioneered vernacular architecture studies in the northeastern United States. For New England, Cummings (1984), Lewandoskie (1993), Stachiw and Small (1989), Steinitz (1988, 1989), and St. George (1982) have documented a wide range of vernacular house forms. Addressing social agency within vernacular architectural contexts, Garrison (1991), Herman (1978, 1984, 1992), St. George (1986), and Sweeney (1984, 1993), among others, have developed sensitive models.
2. Outside Boulton's studies on slavery (1985, 1991), the spatial arrangements of New England household service have received little direct study. The pioneering architectural histories of New England houses—for example, Kelley (1924, 1931), Kimball (1922), Isham and Brown (1965), Major (1926), and Morrison (1952)—systematically disregard the service sector.
3. Dennett (1995) notes the danger of using the architectural metaphor of "spandrels" to underpin an already controversial concept (exaptation) in evolutionary biology. Dennett also charges that Gould has misinterpreted the nature of pendentives in architecture. Nevertheless, Mark (1996) finds at least loose justification for Gould's architectural interpretation, on the basis of historical contingencies in the development of the technology for European classical and neoclassical buildings. Ironically, although the

"spandrel" example may be justified in technological contexts, its appropriateness for biological systems remains open to question.

REFERENCES CITED

Barley, M. W.
1987 *The English Farmhouse and Cottage.* Alan Sutton, Gloucester.

Bourdieu, P.
1977 *Outline of a Theory of Practice.* Translated by R. Nice. Cambridge University Press, Cambridge.
1984 *Distinction: A Social Critique of the Judgment of Taste.* Translated by R. Nice. Harvard University Press, Cambridge, Mass.

Boulton, A. O.
1985 New England's Slave Quarters. *Journal of Regional Cultures* 4/5:57–74.
1991 *The Architecture of Slavery.* Ph.D. dissertation, Department of History, College of William and Mary, Williamsburg, Va. University Microfilms, Ann Arbor.

Brunskill, R. W.
1970 *Illustrated Handbook of Vernacular Architecture.* Faber and Faber, London.

Cummings, A. L.
1984 *The Framed Houses of Massachusetts Bay.* Belknap Press, Boston.

Dennett, D. C.
1995 *Darwin's Dangerous Idea: Evolution and the Meaning of Life.* Simon & Schuster, New York.

Dobres, M-A., and C. R. Hoffman
1994 Social Agency and the Dynamics of Prehistoric Technology. *Journal of Archaeological Method and Theory* 1(3):211–258.

Fischer, D. H.
1989 *Albion's Seed: Four British Folkways in America.* Oxford University Press, New York.

Galenson, D. W.
1981 *White Servitude in Colonial America: An Economic Analysis.* Cambridge University Press, New York.

Garrison, J. R.
1991 *Landscape and Material Life in Franklin County, Massachusetts, 1770–1860.* University of Tennessee Press, Knoxville.

Giddens, A.
1979 *Central Problems in Social Theory: Action, Structure, and Contradiction in Social Analysis.* University of California Press, Berkeley.
1984 *The Constitution of Society: Outline of the Theory of Structuration.* University of California Press, Berkeley.

Glassie, H. H.

1968 *Pattern in the Material Folk Culture of the Eastern United States.* University of Pennsylvania Press, Philadelphia.

1975 *Folk Housing in Middle Virginia.* University of Tennessee Press, Knoxville.

Gould, S. J.

1991 Exaptation: A Crucial Tool for an Evolutionary Psychology. *Journal of Social Issues* 47:43–65.

1993 Fulfilling the Spandrels of the World and Mind. In *Understanding Scientific Prose,* edited by J. Selzer, pp. 310–336. University of Wisconsin Press, Madison.

Gould, S. J., and R. C. Lewontin

1979 The Spandrels of San Marco and the Panglossian Paradigm: A Critique of the Adaptationist Programme. *Proceedings of the Royal Society, London* B 205:581–598.

Gould, S. J., and E. S. Vrba

1982 Exaptation—A Missing Term in the Science of Form. *Paleobiology* 8(1):4–15.

Herman, B. L.

1978 Continuity and Change in Traditional Architecture: The Continental Plan Farmhouse in Middle North Carolina. In *Carolina Dwelling, Towards Preservation of Place: In Celebration of the Vernacular Landscape,* edited by D. Swaim, pp. 160–171. Student Publication of the School of Design, North Carolina State University, Raleigh.

1984 Multiple Materials, Multiple Meanings: The Fortunes of Thomas Mendenhall. *Winterthur Portfolio* 19(1):67–86.

1992 *The Stolen House.* University of Virginia Press, Charlottesville.

Hubka, T. C.

1984 *Big House, Little House, Back House, Barn: The Connected Farm Buildings of New England.* University Press of New England, Hanover, N.H.

1986 The New England Farmhouse Ell: Fact and Symbol of Nineteenth-Century Farm Improvement. In *Perspectives in Vernacular Architecture II,* edited by C. Wells, pp. 161–166. University of Missouri Press, Colombia.

Hughes, T. P.

1990 From Deterministic Dynamos to Seamless-Web Systems. In *Engineering as a Social Enterprise,* edited by H. Sladovich, pp. 7–25. National Academy Press, Washington, D.C.

Isham, N. M., and A. F. Brown

1965 *Early Connecticut Houses: An Historical and Architectural Study.* Republished by Dover Publications, New York.

Izard, H.

1993 The Ward Family and Their "Helps": Domestic Work, Workers, and Relationships on a New England Farm, 1787–1866. *Proceedings of the American Antiquarian Society* 103(1):61–90.

1996 *Another Place in Time: The Material and Social Worlds of Sturbridge, Massachusetts, from Settlement to 1850*. Unpublished Ph.D. dissertation, Graduate School of Arts and Sciences, Boston University, Boston.

Johnson, M. H.

1989 Conceptions of Agency in Archaeological Interpretation. *Journal of Anthropological Archaeology* 8:189–211.

1993 *Housing Culture: Traditional Architecture in an English Landscape*. Smithsonian Institution Press, Washington, D.C.

Kelly, J. F.

1924 *Early Domestic Architecture of Connecticut*. Yale University Press, New Haven, Conn.

1931 *Early Connecticut Architecture*. William Helburn, New York.

Kimball, F.

1922 *Domestic Architecture of the American Colonies and of the Early Republic*. Charles Scribner's Sons, New York.

Kniffen, F. B.

1986 Folk Housing: Key to Diffusion. Reprinted in *Common Places: Readings in Vernacular Architecture*, edited by D. Upton and J. M. Vlach, pp. 3–27. University of Georgia Press, Athens.

Kussmaul, A.

1981 *Servants in Husbandry in Early Modern England*. Cambridge University Press, Cambridge.

Larick, R.

1985 Spears, Style, and Time among Maa-Speaking Pastoralists. *Journal of Anthropological Archaeology* 4:206–220.

1986 Age Grading and Ethnicity in the Style of Loikop (Samburu) Spears. *World Archaeology* 18(2):268–282.

1991 Warriors and Blacksmiths: Mediating Style in East African Spears. *Journal of Anthropological Archaeology* 10(3):299–331.

Laslett, P.

1965 *The World We Have Lost*. Methuen, London.

Lasser, C. S.

1982 *Mistress, Maid, and Market: The Transformation of Domestic Service in New England, 1790–1870*. Ph.D. dissertation, Department of American Studies, Harvard University. University Microfilms, Ann Arbor.

Lechtman, H.

1977 Style in Technology: Some Early Thoughts. In *Material Culture: Styles, Organization, and Dynamics of Technology*, edited by H. Lechtman and R. S. Merrill, pp. 3–20. West Publishing Co., St. Paul, Minn.

Lewandoskie, J. L.

1993 The Early House in Northeastern Vermont: Typical and Atypical Forms, 1770–1830. *Vermont History* 61(1):28–30.

McMurry, S. A.

1984 Progressive Farm Families and Their Houses, 1830–1855: A Study in Independent Design. *Agricultural History* 58(3):330–346.

1988 *Families and Farmhouses in Nineteenth-Century America: Vernacular Design and Social Change.* Oxford University Press, New York.

Major, H.

1926 *The Domestic Architecture of the Early American Republic: The Greek Revival.* Lippincott, Philadelphia.

Mark, R.

1996 Architecture and Evolution. *American Scientist* 84(4):383–389.

Morrison, H.

1952 *Early American Architecture.* Oxford University Press, New York.

Nylander, J. C.

1993 *Our Own Snug Fireside: Images of the New England Home, 1760–1860.* Alfred A. Knopf, New York.

Ortner, S. B.

1984 Theory in Anthropology since the Sixties. *Comparative Studies in Society and History* 26(1):126–166.

Pfaffenberger, B.

1992 Social Anthropology of Technology. *Annual Review of Anthropology* 21:491–516.

St. George, R. B.

1982 *A Retreat from the Wilderness: Pattern in the Domestic Environments of Southeastern New England, 1630–1730.* Ph.D. dissertation, Department of Folklore, University of Pennsylvania, Philadelphia. University Microfilms, Ann Arbor.

1986 "Set Thine House in Order:" The Domestication of the Yeomanry in Seventeenth-Century New England. In *Common Places: Readings in American Vernacular Architecture,* edited by D. Upton and J. M. Vlach, pp. 336–364. University of Georgia Press, Athens.

Stachiw, M., and N. Small

1989 Tradition and Transformation: Rural Architectural Change in Nineteenth-Century Central Massachusetts. In *Perspectives in Vernacular Architecture III,* edited by T. Carter and B. L. Herman, pp. 135–148. University of Missouri Press, Colombia.

Steinitz, M.

1988 *Landmark and Shelter: Domestic Architecture in the Cultural Landscape of the Central Uplands of Massachusetts in the Eighteenth Century.* Ph.D. dissertation, Department of Geography, Clark University, Worcester, Mass. University Microfilms, Ann Arbor.

1989 Rethinking Geographic Approaches to the Common House: The Evidence from Eighteenth-Century Massachusetts. In *Perspectives in Vernacular Architecture III,* edited by T. Carter and B. L. Herman, pp. 16–26. University of Missouri Press, Colombia.

Sweeney, K. M.

1984 Mansion People: Kinship, Class, and Architecture in Western Massachusetts in the Mid-Eighteenth Century. *Winterthur Portfolio* 19(4):231–255.

1993 Meetinghouses, Town Houses, and Churches: Changing Perceptions of Sacred and Secular Space in Southern New England, 1720–1850. *Winterthur Portfolio* 28(1):59–93.

Ulrich, L. T.

1982 *Good Wives: Image and Reality in the Lives of Women in Northern New England, 1650–1750.* Alfred A. Knopf, New York.

1990 *A Midwife's Tale: The Life of Martha Ballard, Based on Her Diary, 1785–1812.* Vintage Books, New York.

Wobst, H. M.

1977 Stylistic Behavior and Information Exchange. In *Papers for the Director: Research Essays in Honor of James B. Griffin,* edited by C. E. Cleland, pp. 317–342. Anthropological Papers No. 61. University of Michigan, Museum of Anthropology, Ann Arbor.

Wrightson, K.

1982 *English Society, 1580–1680.* Hutchinson, London.

II TECHNOLOGICAL POLITICS

5. Intentional Damage as Technological Agency: Breaking Metals in Late Prehistoric Mallorca, Spain

CHRISTOPHER R. HOFFMAN

A SINGULARLY BROAD, ALMOST BOUNDLESS, DEFINITION OF TECHNOLOGY is being put forward by the contributors to this volume, one that encompasses more than materials, tools, and technical knowledge, but some combination of these and a fair dose of gestures, social action, and cultural knowledge. Without all these components, technology is reduced to hardware (Lechtman 1993). However, technology as a concept can be unpacked even further. For example, it seems a fair characterization to say that technology is somehow productive. After all, technologies are used to produce things—to make material culture, and in the context of this volume to make culture as well. The premise of this chapter is that techniques and technical knowledge are also used to break things, even things made by people. This is true in many different cultural contexts, whether modern or ancient, but as is the case with making, breaking is an extremely broad cultural and material phenomenon that can be interpreted from several perspectives and has multiple levels of meanings for those social actors involved. As such, it merits a broad analysis that pays attention to scale, materiality, context, and social agency (Dobres and Hoffman 1994). Breaking is as much about technological practice, politics, and world views as is making. I draw out some of these issues by discussing examples of intentional damage with which I have had personal experience in my own cultural milieu. I then describe intentionally damaged metal artifacts from Copper Age, Bronze Age, and Iron Age sites in Mallorca, Spain. Taken together, these personal and prehistoric examples of intentional damage reveal the complex relationships that acts of destruction implicate among material culture, technology, meaning, power, and social action.

FROM WINE GLASSES TO WARFARE

Glass is a brittle anisotropic material. When produced in thin sheets, it is especially fragile. We know this very well, and we take great pains to keep our glass items, such as wine glasses, from breaking. Shipping and packing companies employ their own tools and bodies of technological knowledge to prevent glass from breaking. Still, almost every item of glass is destined to break eventually. But sometimes people break glass objects on purpose. On certain kinds of occasions (at least in contemporary Western society), people shatter wine glasses intentionally by hurling them into a fireplace. This act has become a metaphor—an almost comical one nowadays—for opulence, display, romance, and important moments.

What is "technological" about throwing a wine glass into a fireplace? Imagine yourself performing this act of wanton wine glass breaking. As you pick up the glass, you quickly judge by its weight, the thickness of the vessel's walls, the distance to the fireplace and other factors how hard to throw the glass and along what trajectory. You probably determined quickly that the item was not one of those plastic imitations. In a split second, you have applied technical knowledge about the material of glass to this specific wine glass and plotted a trajectory for it. The gesture of the throw is, as Mauss would argue, a technical act as well (Schlanger 1990; also Dobres chapter 6 in this volume). As in a dance, you coordinate the parts of your body that will be involved—maybe just your hand and arm but perhaps your whole body—to hurl, toss, or flip the wine glass along the proper trajectory. What is the result? Giddily you anticipate the event as the glass spins out of control toward its destination. Whether or not you hit the desired target (the fireplace), the result is most likely spectacular. Even though you intended this, it still surprises you—the sound of shattered glass, the explosion of shards as the fracture travels through the item in the blink of an eye. At some level, this is dangerous. You could be injured. If you were a child, someone would probably reprimand you. This experiential component of intentional breaking is important—the noise, the mess, the unpredictability, the transgression.

Wine glasses are also intentionally broken toward the end of the Jewish wedding ceremony. The same basic suite of knowledge about the fragile nature of glass and wine glasses, in particular, is applied. After the bride and groom drink wine from a wine glass, the rabbi ceremoniously wraps the glass in a piece of linen and places it on the floor. Instead of throwing the glass against a hard surface, the groom brings his foot down forcefully on the glass. In this case, the act is not only culturally sanctioned; it is an expression of good luck for the newlyweds. Rather than being an act of transgression, it is a vital component of ritual. Beyond that, the act is also a ritually powerful one that constructs new relationships (between men and women and their families) and identities (of husbands, wives, and in-laws). I was at a Jewish wedding recently where the bride and groom secretly substituted two lightbulbs for the wine glass. Because the inside of lightbulbs is for all practical purposes a vacuum, they make a louder noise when crushed. According to my informants, the extremely thin glass shell of the lightbulb also poses fewer safety hazards. The unintended consequences of stepping on glass can be quite dangerous, a piece of technological lore derived from the significant glass knowledge base we possess. Because the lightbulbs were wrapped in a cloth, my wife and I were among the few to know that the groom had not stepped on a wine glass, and it was clear to us that the newlyweds did not want to advertise the fact that they had strayed from tradition.[1]

What frames both of these examples of wine glass destruction is a host of cultural, historical, and personal factors: the goals, opinions, and feelings of those people involved. The acts make sense only within these bounds, and they rely heavily on context. If the techniques used and contexts within which wine glasses were broken in these two examples were switched, their meanings and therefore their impact would be somewhat ambiguous. Taken a step further, if these technological acts were performed in contexts where the breaking of wine glasses normally does not occur (for example, in a hospital waiting room), it might engender any number of reactions, from bewilderment to outrage. This dramatic component reveals social and material action and agency as cultural performance, a cultural performance in which meanings are constructed and interpreted in different ways by different people.

The material dimension of wine glass breaking is a vital component of these acts. The physical properties of glass as a material and the form of the object (a wine glass) as a volume of that material place significant constraints on the amount of force required and the kinds of human gestures that could fracture the artifact. As mentioned previously, the way that glass breaks is vital to understanding the action: it shatters with a relatively loud noise and explosion of pieces. No doubt the crushing of a plastic cup would have a very different effect. Glass breaking, as with many cases of intentional damage, is quintessentially final. The pieces can be recycled, but the originally whole object is no more. More than a starting point for human interpretation, the materiality of the items and these acts of destruction are layered into the construction of meaning.

Of course, the functions, meanings, and symbolism of the original unbroken object are also crucial. In this case, the original object is a wine glass, an object used during celebrations to hold a special consumable liquid. At the end of the Jewish wedding ceremony, a pair of lightbulbs can be substituted for a wine glass precisely because they are hidden from view by the linen wrapping. In the example involving the fireplace, the substitution of a lightbulb simply would not work.

Like many cultural phenomena, the intentional damage of material culture is extremely broad and difficult to define and categorize. It can be seen in a wide variety of social contexts, produces a variety of results, and is directed toward a variety of audiences. In fact, it is certainly as broad as more typically "productive" technological acts. Compare the previous examples of wine glass breaking with the tools and techniques of modern warfare. The design, construction, and use of the tools of modern warfare are dramatically cultural and meaningful at different levels to different people. Needless to say, they are contested and socially divisive as well. Consider the efforts taken to design and construct the first atomic bombs and the subsequent social consequences felt around the world. Dropped from the "wombs" of airplanes, bombs have long been referred to as "babies," their goal to give birth to peace through the act (or threat) of destruction (Dobres, personal communication).[2] Interestingly, the logic behind the cold war military buildup (and the continued development of weapons of destruction) is that it would provide a deterrent to aggression. That is, a weapon of destruction is used to *construct* peace. Even if the bomb is never used, however, it is the potential for the destruction of life, environment, and material culture that makes it a deterrent. Materially and meaningfully, destruction and construction are dialectically linked.

From a technological standpoint, a bomb is also very different from a wine glass. First, from the perspective of the designer, the bomb is intended for destruction (of itself and of its target) whereas the wine glass was designed to hold liquid. Second, the scale and organization of the technological effort put into the design, production, and use of bombs are wholly different from the effort put into designing and producing wine glasses. At another level, though, throwing a wine glass into a fireplace and dropping a bomb are similar: they are both culturally sanctioned acts. Other examples of intentional damage or destruction are not. During various political movements of the twentieth century in the United States, fire and burning have been used to damage or destroy material culture in displays that did not receive the approval of those in power. Women burned their bras, men burned their draft cards, and inhabitants of inner-city ghettoes burned their cities. The symbolic destruction of these made things was as powerful a statement as the actual loss of property, profits, and labor. These acts helped engender a view that there was also a loss of control.

This, then, could become an important way to differentiate between such technological acts of intentional damage. Breaking made things can invoke a wide range of social metaphors, some of them constructive in the sense that they reinforce some social norm or structure, others, acts of social deconstruction that question or critique. Though heuristically useful, this neat distinction between breaking as constructive and deconstructive is complicated by the multiplicity of meanings and interpretations made by people, especially in the examples from modern society mentioned above. Not surprisingly, the meanings of breaking can be simultaneously constructive and deconstructive.

The meanings and messages of these acts of intentional damage, even in supposedly critical deconstructive acts, can only make sense for social actors—can only be deciphered by them—if they work within an existing system of metaphors and structures of social action. In this sense, the material, technological, and experiential components of the destructive act are very important. The act is successful in the sense of making meaning partly by juxtaposing normally disparate materials and techniques (such as a bras and fire), particularly for those deconstructive acts that critique or question normative values. But because the act of intentional damage plays on the same metaphors (as in gender construction and power), it can serve to reify more deeply rooted structures of social action.

Practice theories of social action and agency provide the linkages needed to situate these acts and their multiple and sometimes conflicting meanings in relation to ongoing structures (Bourdieu 1977; Giddens 1984; Ortner 1984). Acts of breaking, like acts of making, can simultaneously maintain tradition and open up possibilities for change. As Dobres discusses in chapter 6, the very gestures and body language employed by individuals during technological performances hold meanings about identities and help form a field of antecedent conditions. During acts of breaking, unintended consequences might be particularly relevant.

VANDALISM AND GRAFFITI: PERSONAL EXPERIENCES

When I was a young teenager, my male friends and I participated in coming-of-age rituals involving destructive events referred to, by us as well as our victims, as "vandalism." At first, our activities were mild: we threw eggs at houses and enveloped cars or trees in toilet paper. Within a few years, as testosterone pumped into our bodies and the social ladder of high school culture in suburban Colorado became increasingly competitive and rigid, our Saturday night excursions turned to beer consumption and more destructive events. We blew up mailboxes and cut holes in fences.

In carrying out these tasks, we developed and employed skills, tools, and a knowledge of materials. None of these components were very specialized, but they were technical nevertheless. Our bodies were our main tools, but we also took advantage of the power of fireworks and bolt cutters. From a generation of kids that preceded us, we learned how to juxtapose everyday items, such as eggs and toilet paper, with other everyday items, such as houses and automobiles, into combinations that were perhaps only mildly damaging but significantly annoying to our victims. More than this, however, these were collaborative technological efforts: by operating as a group, we were able to solve problems, wreak havoc, and generally encourage each other to take part. We built social relationships, constructing a seemingly close-knit group of young men bonded together by dissatisfaction and vandalism.

For us, these acts were part of becoming men and therefore of constructing our gender identities. We rebelled. We tested and resisted structures of authority. We competed with one another for a form of prestige that was destructive in many ways. Not surprisingly, we developed codes of acceptable and unacceptable behavior that were different for young men and young women. These acts of vandalism were about power, something we lacked socially. We could not vote, drive, drink (at least legally), or question what we saw to be a flawed society in any officially sanctioned ways. But we had physical power and the courage and creativity inspired by our competitive cheap beer feasting. By breaking things, we gained power—as if the object itself contained a social energy imparted to it by its maker and its owner.

Certainly these were acts of social deconstruction; we were dissatisfied and disenchanted with the world around us, so we drank beer and broke things. No doubt we were not very skilled at articulating our dissatisfaction in constructive ways, but we did talk about the things in the world that angered us: pollution, soap operas, politicians, and laws and rules that controlled our actions. At the same time, we also reinforced social structures that are still very strong in our culture. One in particular is known as the "generation gap." We were aware that earlier generations of adolescents had experienced the same anger, and their lack of power as well as their current membership in the dominant power generation was a source of significant frustration for us. We might have guessed that through these acts of material destruction and social deconstruction, a generation of soon-to-be-adult-men was being constructed. Certainly these acts helped shape feelings and opinions I later came to hold about material culture, particularly material culture that belonged to me, that was my "property."

Consider another example from modern urban and suburban North American culture: graffiti.[3] Berkeley (California), the place where I live and work, has graffiti all over it—on buses, on signposts, on trees, buildings, and backpacks. Like a blanket, the colors, letters, and variety of styles form a nearly indecipherable collage over the city. Though I am beginning to gain some sort of appreciation for graffiti, I have not always been so understanding. In fact, despite my own background (or probably because of it), I have over the years developed strong feelings about respect for public and private property. Until quite recently, therefore, I saw graffiti only as a form of vandalism, an act of defacement that invaded a space that was public and communally shared, and therefore to a degree mine.

Property owners, local government officials, and other powerful social entities no doubt hold a similar view. Despite a brief love affair with New York City's painted trains, graffiti is largely seen to be a sign of vandalism, urban decay, and gang activity.[4] Cities such as San Francisco and New York spend enormous sums "battling the graffiti problem," and increasingly severe penalties are levied against those caught. There is a sense of competition between writers (as graffiti artists refer to themselves) and the public at large: it is common to find a tag (a writer's unique graffiti signature) written on top of a square of paint that covers an older tag. New exterior materials—tile and paint—are developed that resist graffiti or are easily cleaned, but innovative solutions are found to mark even these: stickers are applied, tile grout is marked up, and nearby objects, such as trees, become hosts for signs. Public transportation agencies hire armies to clean buses after hours, so writers resort to more indelible markings (tags incised with a blade, or holes burned in using lighters). Given that graffiti writers work 24 hours a day, there seems to be little hope that these official approaches will be effective.

More recently, a number of somewhat more innovative responses to the graffiti "problem" have been developed. Some of these are social, in the sense that they target the graffiti writer. Within this category of responses, the official representation of the writer is an important variable. For example, where the graffiti writer is viewed as a criminal, the sentence is to have those convicted of the crime clean graffiti.[5] Other views treat the graffiti writer as an individual trying to express his or her art: these writers are given the option of joining public beautification projects in which they can express their art in publicly sanctioned forms. Other responses to graffiti look for solutions in sophisticated technologies: spectrophotometer color-matching systems (to ensure that the paint used to cover tags matches the original surface color) and advanced surveillance methods.

One problem limiting most responses to graffiti is the public perception that it is a form of vandalism perpetrated by gang members to mark territory. Graffiti is therefore treated as if it were a gang-related crime. Though gang graffiti does happen, it probably explains only a fraction of graffiti cases. In fact, young people "tag" things such as buses, signs, phone booths, and backpacks for a variety of reasons, not the least of which is the identification and advertisement of the self and the thrill resulting from the risk taken (Ong 1990). Within the larger context of American society, the multiplicity of meanings inherent in graffiti is astounding.

Regardless of its legal status, graffiti is the product of technological practice, a complex interweaving of materials, knowledge, procedures, and meanings. The writer practices his or her technique sometimes over a period of several years and learns from other writers. A sophisticated toolkit also plays an important role in the graffiti writers' experience. Specially designed tips are stuck on spray paint cans to provide the special effects seen in works of graffiti. In fact, the standard tip that comes on a can of spray paint is derogatorily referred to as a "sucker tip." Although the sale to minors of certain kinds of markers and spray paint is now illegal, these tools are still employed to produce the vast majority of tags. Because writers take pride in stealing their materials, making their sale to minors illegal could have only a minimal impact. No doubt a sophisticated network of relationships exists to provide the raw materials and tools needed by the graffiti artist.

Even under cursory inspection, it is evident that writers have distinctive styles, whether the piece is text only, figurative, or some combination of the two.

Traveling along the highway, one can follow the work of individual writers from one overpass to the next. Sitting in the back of the bus one day, I gained some unexpected insight into this dimension. Two young men were sitting across from me. It was a school day, around 10:00 in the morning. Knowing that it is difficult to keep young people in classrooms these days, I was not surprised to see them. Both had that practiced tough look that other people on buses like to stay away from. One pulled a notebook out of his backpack and leafing through to a selected page, showed the other what was clearly an outline of a graffiti piece (the kind with big bulging letters) that they intended to write somewhere. As he pointed out some detail of the piece, I could glimpse other outlines on other pages. The other individual noticed me looking at the book and motioned to his companion to put the book away. No doubt such items are not intended for my public consumption. I later learned that these notebooks, called "piecebooks" or "blackbooks," are widely used by graffiti artists and that the outlines they contain are occasionally exchanged by artists.

The interaction of these two young men reveals the collaborative nature of graffiti. Though writers do occasionally work alone, more frequently they are part of a "crew," a group that works together to make graffiti. As was true in the vandalism example discussed previously, through cooperation and contestation, graffiti writers construct themselves and their culture, and the act of "putting up" graffiti is an important moment of creation. As product and process, for those participating graffiti is an important part of youth culture. Its language is part of youth culture, as is its style and expressive form. Within the graffiti writer's community, furthermore, there are recognized levels of knowledge and a stated code honoring and respecting more senior writers. Graffiti is used by these people to define their world and their relationships to one another.

Though graffiti is normally seen as a male form of expression, I have seen young women tagging phone booths, and there are well-known female graffiti artists. Furthermore, I suggest that young women might be responsible for a large amount of tagging of more portable items: backpacks, shoes, and notebooks.

What motivates graffiti writers to take pen in hand? Walter Ong, an authority on the culture of writing and literacy, makes the compelling case that graffiti is about "self seeking recognition, as all selves do" (Ong 1990:406). Its expression as graffiti, he argues, relies as much on an outlaw mentality as it does a different perspective on urban beautification and display advertising. For Ong, graffiti is a "high-tech" response to a high-tech culture that dehumanizes and depersonalizes the individual.

I overheard a group of young people on the bus one day. One guy said, "Yeah, this bus is just waiting to be painted. It's so white." (By the way, these were all white youths from relatively affluent neighborhoods.) The bus in question, however, was not white at all. Rather, "white" referred to the sterile, institutional, and somewhat boring environment of the bus. In addition to rebelling, testing authority structures, and competing with each other, these young people are promoting their own aesthetic sensibility, something I have to admit I was not too concerned with when I was their age. A quick survey of fashion and art demonstrates how successful they have been in this effort, but whether this is an intended consequence or overtly conscious is not clear.

This analysis of graffiti has been limited to personal observations and examples. I cannot help but think that they provide too shallow a picture of the complexity of meanings, actions, and technical concerns implicated in graffiti, of which only one small part is the end-product. Although it is well known that graffiti writers interpret their work differently from the police, it is probably also true that within the graffiti-writing community there are significantly different interpretations and explanations. These differences no doubt fall along a number of lines, gender and skill level coming to mind quickly. In the literature I surveyed, the writers who were interviewed appeared to represent the opinions of senior writers, and no doubt their position in the graffiti culture colors their perception. For example, they largely downplayed the role of gang graffiti and seemed to imply that the goal of young inexperienced writers (known as "toyz") was to attain the kind of mastery that they had, probably under the mentorship of another. I wonder if writers from gangs and the young people from the high school near me would fully agree with these assessments.

BREAKING METALS IN LATE PREHISTORIC MALLORCA, SPAIN

The preceding examples of graffiti and vandalism were drawn from personal observations in two modern urban/suburban areas in the western United States (Denver, Colorado, and Berkeley, California). They are further limited because they relate to a very specific form of intentional damage: that performed by dissatisfied and disenfranchised youths in suburban and urban areas. However, there are numerous examples from other archaeological, anthropological, and historical contexts that reveal the broad nature of this phenomenon. Graffiti is abundant on the walls of Pompeii and other Roman cities (Brilliant 1979). Though much of this writing on walls appears to have been more culturally sanctioned (sensu, political endorsements) than the modern counterparts previously discussed, other inscriptions include humorous notes, greetings, and declarations of love (Cole 1931; Della Corte 1927). Quipped one writer, "I am surprised, O wall, that you have not fallen down under the burden of so many tedious writers." Fausto Niccolini (1854–1896) compiled a large series of graffiti inscriptions from Pompeii, in-

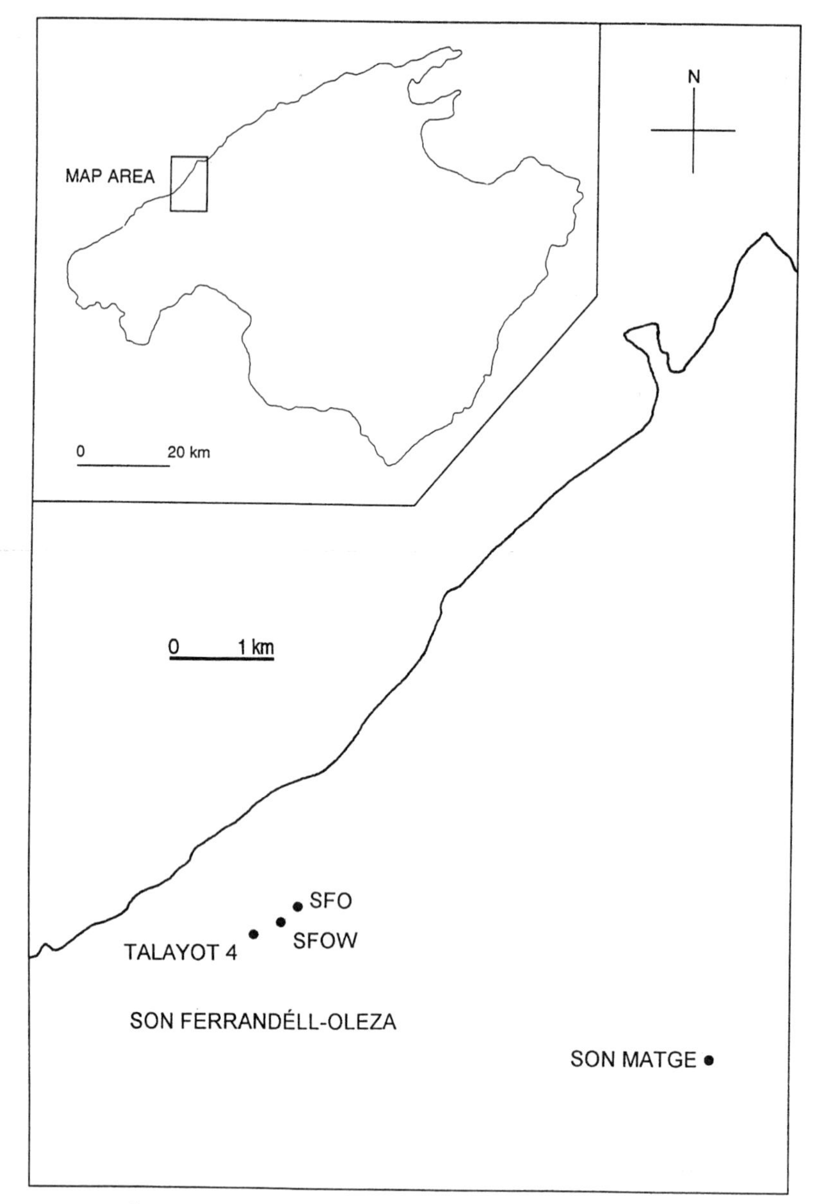

Figure 5.1. Mallorca (Balearic Islands, Spain).

cluding examples of "gladiatorial graffiti"—figures and text drawn by gladiators on colosseum walls (in Brilliant 1979). In addition, there are numerous examples from the Near East and the New World of invading peoples vandalizing political and religious symbols (especially on statues and churches) in a symbolic act of conquering and desecration.

In fact, it was an archaeological example of this phenomenon that drew me to the general cross-cultural topic of intentional damage. During my dissertation research on metals and metallurgy in late prehistoric Mallorca, Spain (Hoffman 1991a, 1993, 1995), I encountered metal artifacts that were intentionally broken, and yet it was clear that the motivations behind and contexts of these events of breaking were different in each of the three time periods I studied: the Copper Age, the Bronze Age, and the Iron Age. Because of my interest in the entire "life history" of the artifact—from the procurement of its raw materials to its deposition in the ground—it was necessary for me to think about the meanings, functions, and technologies of these acts of breaking. I argue that these broken artifacts are the outcome of purposeful, motivated acts that were politically and socially charged and occurred within a meaningful cultural context. Furthermore, the degree and kind of technical knowledge employed helped shape the meanings of those acts.

Despite a long history of archaeological investigations in the Balearic Islands (figure 5.1), there are still significant shortcomings in the general understanding of social structure, economy, and even basic culture history. However, a series of excavations over the past 25 years or so in a circumscribed area on the northwest coast have opened up interesting analytical and interpretive possibilities (Waldren 1982, 1986). In this area, Waldren and colleagues have surveyed a wide area and conducted excavations at dozens of sites. In my dissertation, I had the opportunity to examine nearly all the metal artifacts from these excavations. This assemblage included intentionally damaged artifacts from three time periods—the Copper Age, the Bronze Age, and the Iron Age.

"ONE HALF FOR YOU AND THE OTHER FOR ME": DIVIDING AN INGOT IN THE COPPER AGE

Though evidence exists for human presence on Mallorca as early as 5000 B.C., the Mallorcan Pretalayotic Copper Age (2000–1300 radiocarbon B.C.) was a period of significantly higher population levels and more permanent habitation than earlier periods. Chronologically, the Mallorcan Pretalayotic overlaps with the Spanish Chalcolithic and Early to Middle Bronze Ages. Though of a different scale, the basic economic and social changes evident in the archaeological record of Mallorca are similar to developments in Spain and other parts of western Mediterranean Europe (Chapman 1990). During this period, the stylistically contentious Bell Beaker ceramic was in circulation, and metallurgy was invented independently in southeastern Spain. In Mallorca, habitation in caves and rockshelters was relatively common, as was habitation and collective inhumation in artificially enlarged caves and megalithic structures called *navetas,* or *navetiformes.* The open-air

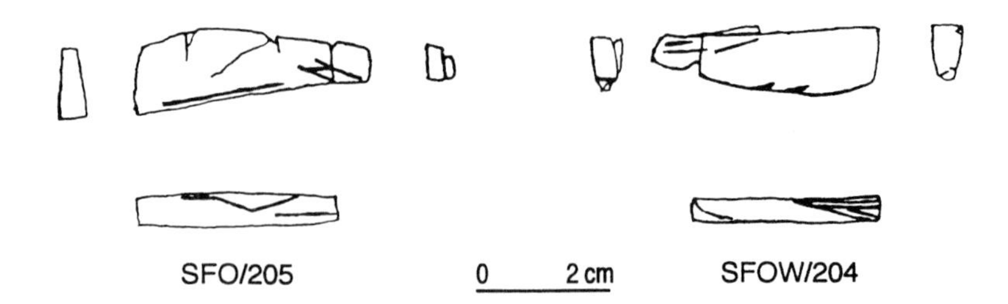

Figure 5.2. Two pieces from shared Pretalayotic worked blank.

walled settlement of Son Ferrandéll-Oleza has been under excavation for several years (Waldren 1987).

Within the study area of the dissertation, ceramic crucibles and stone flat molds were used to cast artifacts of pure copper and alloys of copper with small additions of tin or arsenic, or both. Macroscopic and microscopic metallographic examination of these objects indicates that they were heavily forged and incompletely annealed.[6] Finished artifacts were generally simple types (such as awls and flat daggers) and unique utilitarian items that lacked typologically diagnostic features. The overall impression of Pretalayotic metal artifacts and their production is one of variability, versatility, and nonstandardization (Hoffman 1993).

During the occupation of the open-air settlements of Son Ferrandéll-Oleza (SFO and SFOW), a rectangular artifact of nearly pure copper was broken in half (figure 5.2). Though this item was originally identified as an ingot, it has been partially cold-worked, a condition that led Peter Northover to label it a "partially worked blank." The piece was first struck several times along its long axis with a sharp (presumably metal) object having a chisel end (approximately 0.5 centimeters wide). Each blow penetrated into the metal approximately half a millimeter. The same action was taken on the opposite side.[7] At one end, successive blows tore through to the other side of the blank. Apparently this crack was used to pull the object apart (though the thick patina makes it difficult to determine with certainty how this final separation was accomplished). This action divided the object very accurately into two pieces: one half weighed 9.1 grams, and the other weighed 8.8 grams.

One could argue that, strictly speaking, this is not an example of intentional damage because the act did not "damage" the artifact. Rather, as I discuss next, the motivation behind this act was to divide a piece of semi-raw material for consumption by two parties. On the other hand, the original object was broken; it was made incomplete via an act of force. Specialized tools, energy, and a knowledge of metals were employed to divide this piece. Certainly breaking can be part of a manufacturing or curation process as well. Crucible melting of previously manufactured items was clearly practiced on at least one of the sites where these artifacts were found (SFO). The presence of such melting crucibles on other Mallorcan sites of this period suggests that crucible melting was a prehistoric form of recycling, but this method was not used to distribute the partially worked blank to two parties. Sharing the object, like "breaking bread," was accomplished by breaking.

The deposition context of these two halves supports the hypothesis that this was an act of sharing. One half was deposited at the Old Settlement of Son Ferrandéll-Oleza (SFO); the other at the neighboring site of the West Old Settlement (SFOW), located some 50 meters away. Originally, the excavator identified the West Old Settlement as a dump zone for the Old Settlement, but my analysis of these materials leads me to interpret them as contemporaneously occupied settlements. The presence of these artifacts is one line of evidence. They are clearly part of the same object: their composition and lead isotope ratios are identical, and they can be rejoined along the line of the fracture.

Though it is possible that one half was deposited on a settlement and the other was thrown out with the trash, the contemporary settlement hypothesis makes more sense. The identification of the West Old Settlement as a dump zone was based largely on its lack of stone architecture. However, the artifact assemblage and radiocarbon dates indicate that the West Old Settlement was used during the earlier part of the Old Settlement occupation (2000–1700 radiocarbon B.C.). At that time, the Old Settlement lacked much of its stone architecture as well. Furthermore, the same range of activities (involving the production and use of ceramics, flaked and ground stone, metals, animal bone, and foodstuffs) is evident at both sites, and the materials in the West Old Settlement have a patterned and clustered distribution (Ensenyat Alcover and Waldren 1987; Hoffman 1993).

What was the relationship between the groups occupying the two settlements? It is certainly possible that this pair of sites acted as a single settlement, consisting of two dispersed extended households or corporate groups. On the other hand, they might have acted as relatively distinct socioeconomic units. At this time, the economic data from these sites are still under investigation. Regardless of the specific relationship between the two settlements, the act of splitting a single piece of copper and distributing it between two neighboring sites suggests the sharing of materials and perhaps pooling of resources in order to obtain such materials. The fact that these two pieces were nearly identical in weight is relevant.

Equally interesting, this settlement pattern was not stable in the long term. Sometime around 1700 B.C., the West Old Settlement was abandoned and many of the stone structures of the Old Settlement were constructed. Whether the community of the West Old Settlement moved into the now walled Old Settlement or moved off to some other location is not clear. Apparently the original relationship relied on a tradition of cooperation. Like other parts of western

Mediterranean Europe at this time, on Mallorca there is a significant trend toward vertical hierarchization and social and political competition between and presumably within communities (Chapman 1990).

INTENTIONAL DAMAGE AND TECHNOLOGICAL SOPHISTICATION: A LATE BRONZE AGE "CACHE"

The Late Bronze Age or Talayotic Bronze Age of Mallorca (1300–800 B.C.) overlaps with the Spanish Middle to Late Bronze Age. Though social change in western Mediterranean Europe was relatively stable during this period (when compared with central Europe, for example), in Mallorca and neighboring areas evidence suggests increasing levels of settlement and social hierarchy. The period is named after the stone-built towers (*talayot* or *talaiot*) that dot the Mallorcan landscape. Caves and rockshelters continued to be used (though more for funerary ritual and temporary habitation), and there was a significant move to the level plains of the island for primary occupation and presumably agricultural activities (though few relevant economic data have been collected).

When compared with the earlier period, Talayotic Bronze Age metallurgy was characterized by a higher degree of standardization and technological sophistication, and this is evident throughout the *chaîne opératoire* as well as the remainder of their use-life (including deposition). Objects of metal were produced using a more regular alloy of tin bronze. More sophisticated techniques were employed on a regular basis. Bronzes were produced using complex casting techniques, and they were regularly subjected to complete annealing and a careful final cold-working. Whereas the artifact types circulated in the previous period were morphologically and stylistically generic, many objects in the Talayotic period were typologically distinctive: swords, palstaves, and spearheads having singular characteristics (Delibes de Castro and Fernández-Miranda 1988). At the same time, other than rejuvenation (sharpening) of bronze awls at Son Matge, there is no evidence in the study area for metallurgical production, suggesting that communities here were dependent on other communities for metal objects. Overall, a more structured and standard range of metallurgical products and practices were socially acceptable during this period.

Also during this period, several talayots were constructed in the Son Ferrandéll-Oleza area, adjacent to the Pretalayotic sites. A collection of broken bronze artifacts were found buried under the floor of the entryway to the structure called "Talayot 4" (Chapman and Grant 1989; Chapman et al. 1993). On the surface, it would appear that this collection of functionally useless artifacts could be called a horde of "scrap metal." However, microscopic and macroscopic analysis of the artifacts and their fractures indicates that this attribution could only be a partial explanation, at best, and an incorrect one at worst. The characteristics of the fractures suggest that the artifacts were intentionally damaged using sophisticated metallurgical techniques. One piece is a curved fragment from the middle section of a fairly large sword. The degree of this curve and the twisting of the four corners suggest strongly that the fracturing of this piece was intentional. The second piece is a combination of two artifacts: a socketed axe and the blade of another axe. The socketed axe has been intentionally damaged in three ways. First, the bit has been carefully flattened to a very dull edge. Second, the loop has been bent over. And third, the fractured bit of another axe has been hammered into the socket, effectively rejecting the possibility of attaching a handle. Somebody did not want this object to be used. Because the surface layers have been slightly cold-worked (normally one of the final finishing stages for such artifacts), it was not manufacturing errors that relegated these artifacts to "scrap" status.

The person or persons who performed these acts of intentional damage, particularly in the case of the socketed axe, were technically skilled and understood metallurgical properties. In other words, it was most likely someone who could also manufacture metal artifacts. According to Peter Northover, the dulling of the socketed axe blade was accomplished by careful cold-working and annealing followed by final cold-working. He estimates that the final reduction of this blade was on the order of 30 percent. If not for the other damage done to the artifact, it might appear that this was an intentionally blunt tool of some kind.

The structured depositional context of these broken artifacts, under the floor of the entryway to a talayot, raises some interesting possibilities about the symbolic uses of this deposit. These items were "hidden" from others. Though I think they were buried for later recovery, it is possible that they were originally intended to lie in the entryway forever, as a votive deposit. The fact that before being hidden these objects were damaged using sophisticated technological knowledge and precise force and temperature begs an explanation.

These acts of intentionally damaging metal artifacts and their deposition in the entryway resonate with other Mallorcan Late Bronze Age metallurgical practices. First, the damage done is marked by its technical complexity in the same way that their manufacture was accomplished using a sophisticated and large number of techniques that required specialized technical knowledge and control over the production process. In addition, just as these artifacts were hidden from public view, so, too, their production seems to have been a more hidden, or less widely available, practice (at least in comparison with the earlier period). This is a matter of the social control of technical concerns—of restricting access to the knowledge, skills, and perhaps equipment and raw materials, needed to make these objects. Finally, metallurgical deposition in general during the Talayotic Bronze Age was dominated by structured, overt displays: in burials for instance. (In the earlier period, metal artifacts were found with rubbish.)

Because of this overall resonance, this example of intentional damage may have been socially consistent with the meanings of other metallurgical practices and invoked metaphors of social construction. But this resonance is not unambiguous: if metallurgical production was subject to more explicit social control, then hiding (valuable) semi-raw materials might have been an act of resistance.

DEAD BODIES AND KILLED ARTIFACTS: IRON AGE FUNERARY PRACTICES

The Mallorcan Post Talayotic Iron Age (800–123 B.C.) corresponds to the European Iron Age and is described as a period of increasing interaction with peoples from around the Mediterranean. The variety of site types and material cultures during this period points to a complicated picture of interaction within the context of the Balearic Islands and with groups from elsewhere (Guerrero Ayuso 1984; Hoffman 1991b). According to historical sources, Carthaginians had established a community on the island of Ibiza by 654 B.C. Despite this new element, there are strong signs of continuity with the previous period as well. In metals, objects of iron and lead were in circulation in addition to bronze. Though some were imported to Mallorca, others are unique and suggest local production, though production sites are not known from this period (Delibes de Castro and Fernández-Miranda 1988). Presumably this lack of production evidence is due to the lack of targeted survey and the difficulty involved in locating and identifying production sites. There is also a significant increase in the proportion of decorative items in this period.

During the Mallorcan Post Talayotic Iron Age, the rockshelter of Son Matge was used for a funerary ritual that was inherently destructive. Bodies and grave goods were buried in activated quicklime, and new inhumations were dug into the existing matrix of bodies, artifacts, and lime. Metal artifacts—of iron, bronze, and lead—were relatively common in this matrix, with iron being far more common than either other material. Interestingly, many of the objects recovered from Son Matge's quicklime inhumation deposits were intentionally damaged: ornamental bronze *tintinabula* were hit with heavy objects and crushed, lead plaquettes were bent in half, iron spears were twisted in half and bent into coils. Although this practice is not unique to Mallorca, it does resonate strongly with the destructiveness of the quicklime inhumations. In contrast to the technically sophisticated damage done in the Bronze Age, the damage done here is more of the smashing type. Thus little metallurgical knowledge would have been required, despite the fact that the metallurgical knowledge of the period was sophisticated.

These acts of intentional damage were almost certainly part of the funerary ritual itself and constituted meaningful symbolic acts for those involved in the funerary ritual. They were not intended for recovery. In this case, not only was the act of damaging metal grave goods consistent with the funerary ritual, intentional damage was likely a fundamental component of it.

UNLOCKING THE POWER OF MATERIAL CULTURE THROUGH ACTS OF FORCE AND FINESSE

As a final thought, I return briefly to the issue of technological politics, and of power more broadly. Earlier in the chapter, when I confessed my experiences as a teenage vandal, I described how I gained political stature among my peers and empowered my sense of self by breaking things. For the prehistoric Mallorcan cases discussed here, the metaphor of capturing the social energy of those made things again becomes useful. The acts of intentionally damaging copper blanks, socketed axes, and grave goods effectively bestowed upon the breaker some kind of social prestige or power that could be leveraged for some other purpose.

In making artifacts—whether of paper, stone, or other material—people empower objects with their labor, skill, certain external forces (such as heat and pressure), and the internal potential of the material itself (such as the potential inherent in its crystalline structure or chemical composition). The social and economic relations and identities, world views and meanings, and structures of social action that are equally responsible for the act of creation further make that object a dynamic participant in the contexts of culture. Throughout the remainder of that artifact's "life history," it is embedded in culture, and its meanings are reaffirmed and contested. The end of its use-life may occur in any of several ways: it might simply wear out, be accidentally broken, or lose its appeal and be discarded. On the other hand, it might be deposited in a manner that further connects it to the meaningful construction of culture. The inclusion of material culture in funerary ceremonies is a good example of this phenomenon.

The acts of intentional damage and destruction discussed here (prehistoric and modern) have much in common with this latter kind of structured deposition. After all, they represent intentional acts in which the maintenance and manipulation of meaning occur. Beyond this similarity, intentional damage involves a significant degree of technical knowledge and technological practice. The meanings and power invested into the artifact earlier in its "life history" are again brought into the social sphere. In some cases, a metaphor of social construction is more important to the agent(s) breaking the artifact, whereas in others the person (or persons) performing the damage will be more concerned with turning that history and meaning into an act of subversion and social deconstruction. Despite this intentionality, however, no one person can predict all the consequences of the act, nor can they understand the multiple and sometimes conflicting meanings and interpretations that will result.

The technical knowledge applied in the prehistoric Mallorcan cases becomes important in this regard. In the Copper Age, brute force was used to split the copper ingot. This allowed a piece of semi-raw material, almost certainly something of value, to be shared—a powerful act that might help reinforce the relationship between two groups. By contrast, in the Late Bronze Age, damage was accomplished in a careful and technically sophisticated way. In this case, the human force applied was the technical knowledge of metallurgical practice. This was also a powerful act, but exactly what it enabled is not clear. Here is one hypothesis: because metallurgical knowledge was not widely shared in this period, the act of making this socketed axe nonfunctional effectively guaranteed that whoever had the object would have to bring it to someone skilled in working metal. During the Iron Age, grave goods of iron, bronze, and lead were killed using physical force rather than technical knowledge. Between the violent treatment of the grave goods and the inhumation of the body in activated quicklime, this funerary ritual must have been a spectacular one, brimming with symbolism and meaning.

I take it for granted that these acts were not homogeneously interpreted by members of these communities. They surely had different meanings for, and effects on, children, old people, people with and without metallurgical skills, and teenagers (rebellious or otherwise). That said, I have found it easier to identify how these acts resonated with existing social practices than how they questioned those same practices. Although I do suspect that the maintenance of tradition was more prevalent in these prehistoric Mallorcan societies than in ours, I would not be surprised if at some level I am still constrained by a normative view of prehistoric cultures. But acts of resistance and social criticism are more difficult to identify as well, particularly using archaeological data. I will have to keep my eyes—and mind—open.

What I have tried to show is that these acts of intentional damage were context-specific and embedded within relevant, ongoing structures of social and material action. The breaking of an artifact is just as social an act as the making of that artifact, and in the cases discussed here, intentionality and motivation, that is to say, agency, are facets that cannot be ignored.

NOTES

I thank the following people for assistance in this chapter and the dissertation from which much of it stems: Marcia-Anne Dobres, Ruth Tringham, Meg Conkey, Bill Waldren, Peter Northover, Beth Reingold, and Nathan Meyer. Parts of the dissertation research and writing were supported with assistance from the National Science Foundation, the Center for Accelerator Mass Spectrometry (Lawrence Livermore National Laboratories), a University of California (Berkeley) Fellowship, the Mining and Metallurgical Society of America, the Stahl Endowment (Archaeological Research Facility, University of California, Berkeley), and research funds from the Deia Archaeological Museum and Research Centre.

1. Another informant told me that this substitution of lightbulbs for wine glasses is not uncommon.
2. As discussed by Childs in chapter 2 of this volume, metaphors of childbirth, gender construction, and social reproduction are expressed in African ironworking technologies and technological performances. And Dobres (chapter 6) describes how childbirth (while clearly involving technical knowledge and skill), along with other "domestic" technologies practiced by women, has been culturally devalued in industrial societies and not considered technological per se.
3. In this chapter I am not discussing clever anecdotal graffiti found in bathrooms (known as latrinalia), but tags and larger works normally encountered outside.
4. The source that provided me the best information about the many aspects of graffiti is *Art Crimes: The Writing on the Wall,* an on-line exhibit and graffiti forum located on the World-Wide Web.
5. A somewhat less innovative response along these lines was recently considered by the California State Assembly: paddling as a form of punishment.
6. In the archaeological and archaeometallurgical literature, complete annealing (the removal of microstructural characteristics that could lead to brittleness via the controlled application of high temperature) followed by limited cold-working or forging (hammering to sharpen edges) is assumed to indicate more advanced metallurgical control and sophistication.
7. There are a few chisel marks that are not in line with this axis, but their purpose is not clear.

REFERENCES CITED

Anonymous

1994 *Art Crimes: The Writing on the Wall.* An exhibition of graffiti on the World-Wide Web. http://www.graffiti.org/.

Brilliant, R.

1979 *Pompeii—AD 79.* Clarkson N. Potter, New York.

Bourdieu, P.

1977 *Outline of a Theory of Practice.* Translated by R. Nice. Cambridge University Press, Cambridge.

Chapman, R. W.

1990 *Emerging Complexity: The Later Prehistory of Southeast Spain, Iberia, and the West Mediterranean.* Cambridge University Press, Cambridge.

Chapman, R. W., and A. Grant

1989 The Talayotic Monuments of Mallorca: Formation Processes and Function. *Oxford Journal of Archaeology* 8:55–72.

Chapman, R. W., M. van Strydonck, and W. H. Waldren

1993 Radiocarbon Dating and Talayots: The Example of Son Ferrandéll Oleza. *Antiquity* 67(254):108–116.

Cole, H. W.

1931 *The Writing on the Wall or Glimpses from Pompeian Graffiti into the Daily Life of the Ancient Romans.* Service Bureau for Classical Teachers, New York.

Delibes de Castro, G., and M. Fernández Miranda

1988 *Armas y utensilios de bronce en la Prehistoria de las Baleares.* Studia Archaeológica 78. Universidad de Valladolid, Valladolid.

Della Corte, M

1927 *Pompeii—The New Excavations.* Tipografica di F. Sicignano, Valle di Pompei.

Dobres, M-A., and C. R. Hoffman

1994 Social Agency and the Dynamics of Prehistoric Technology. *Journal of Archaeological Method and Theory* 1(3):211–258.

Ensenyat Alcover, J., and W. H. Waldren

1987 Pottery Distribution Statistics: Ferrandéll-Oleza Copper Age Old Settlement. In *Bell Beakers of the Western Mediterranean,* edited by W. H. Waldren and R.-C. Kennard, pp. 307–368. BAR International Series No. 331(i). British Archaeological Reports, Oxford.

Giddens, A.

1984 *The Constitution of Society: Outline of the Theory of Structuration.* University of California Press, Berkeley.

Guerrero Ayuso, V. M.

1984 *Asentamiento Púnico de Na Guardis.* Excavaciones Arqueológicas en España 133. Ministerio de Cultura, Madrid.

Hoffman, C. R.

1991a The Metals of Son Matge, Mallorca—Technology as Cultural Activity and Behavior. In *IInd Deyá Conference of Prehistory: Archaeological Techniques, Technology, and Theory,* edited by W. H. Waldren, J. Ensenyat, and R. C. Kennard, pp. 169–188. BAR International Series No. 574. Tempus Reparatum, Oxford.

1991b Bronze, Iron and Lead: Iron Age Metallurgy in Mallorca, Spain. In *Recent Trends in Archaeometallurgical Research,* edited by P. D. Glumac, pp. 21–31. MASCA Research Papers in Science and Archaeology No. 8(1). University of Pennsylvania, Philadelphia.

1993 The Social and Technological Dimensions of Copper Age and Bronze Age Metallurgy in Mallorca, Spain. Unpublished Ph.D. dissertation, Department of Anthropology, University of California at Berkeley.

1995 The Making of Material Culture—The Roles of Technology in Late Prehistoric

Iberia. In *Social Complexity in Late Prehistoric Iberia,* edited by K. T. Lillios, pp. 20–31. Archaeological Series No. 8. International Monographs in Prehistory, Ann Arbor.

Lechtman, H.

1993 Technologies of Power: The Andean Case. In *Configurations of Power in Complex Societies,* edited by J. S. Henderson and P. J. Netherly, pp. 244–280. Cornell University Press, Ithaca, N.Y.

Niccolini, F.

1854–1896 *Le case ed i monvmenti di Pompei, disegnati e descritti.* Napoli.

Ong, W. J.

1990 Subway Graffiti and the Design of the Self. In *The State of the Language,* pp. 401–407. Richard Clay, Suffolk.

Ortner, S. B.

1984 Theory in Anthropology since the Sixties. *Comparative Studies in Society and History* 26(1):126–166.

Schlanger, N.

1990 Techniques as Human Action. *Archaeological Review from Cambridge* 9(1):18–26.

Waldren, W. H.

1982 *Balearic Prehistoric Ecology and Culture: The Excavation and Study of Certain Caves, Rock Shelters, and Settlements.* BAR International Series No. 149. British Archaeological Reports, Oxford.

1986 *The Balearic Pentapartite Division of Prehistory.* BAR International Series No. 282. British Archaeological Reports, Oxford.

1987 A Balearic Beaker Model: Ferrandéll-Oleza, Valldemosa, Mallorca, Spain. In *Bell Beakers of the Western Mediterranean,* edited by W. H. Waldren and R. C. Kennard, pp. 207–266. BAR International Series no. 331(i). British Archaeological Reports, Oxford.

6. Technology's Links and *Chaînes*: The Processual Unfolding of Technique and Technician

MARCIA-ANNE DOBRES

ARCHAEOLOGICAL STUDIES OF TECHNOLOGY have been revitalized significantly in recent years by the introduction of an analytic method known as *chaîne opératoire.* André Leroi-Gourhan was the first to name and apply this analytic technique to archaeological inquiry, and today *chaîne opératoire* analysis is employed to detail with extraordinary precision productive sequence(s) and decision-making strategies of raw material transformations, past and present. It is, however, far more than an analytic tool for identifying and describing the material "life history" of artifacts. *Chaîne opératoire* can be a powerful conceptual framework—a methodology—providing technology studies with both the empirical rigor they require and the human face they deserve. In this chapter I argue that *chaîne opératoire* is, at one and the same time, an interpretive methodology and analytic method capable of forging robust inferential links between the material patterning of technical acts and sociopolitical relations of production accounting for them. Although the analytic method of *chaîne opératoire* research is utilized most often to identify the prehistoric mental "maps" (or so-called world views) structuring rule-governed technical activities, as a conceptual methodology (and not simply an analytic method; distinction in Harding 1987), it can be especially helpful in considering the dynamic social milieus and artifice by which material acts were differentially pursued by technicians and variously organized work groups.

This discussion will, I hope, contribute to the revitalization of research on prehistoric technology in two ways. First, it directs attention to some of Mauss's long-forgotten views on technology, in particular his argument that technology is a "total social fact" encompassing more than the material and gestural actions transforming natural resources into cultural products. The next step is to infuse this redefinition with a healthy dose of contemporary social theory, by taking up the question of social agency (especially gender) that is implicated in expressions of self-interest and social identity through technical means. These ideas suggest some ways that the preservable traces of gestural acts of prehistoric artifact production and use can serve as an inferential link to the social agency of the technicians themselves. The discussion is substantiated with reference to a particular social formation of interest to much of archaeological inquiry, the communal mode of production (CMP). Contemporary examples include cases prominent in ethnoarchaeology, such as the !Kung San (Ju/'hoansi), Chipewya, and Iñupiat, and also the Israeli *kibbutz,* which shares relevant structural commonalities. Two archaeological examples from the French Late Upper Palaeolithic round out the discussion.

THE *CHAÎNE OPÉRATOIRE* AND WHAT'S HIDDEN IN BLACK BOXES

To date, *chaîne opératoire* is primarily an analytic tool successfully applied to two fundamental kinds of research questions. The first and most basic is to identify the sequential technical operations by which natural resources were transformed into culturally meaningful and functional objects (Pelegrin et al. 1988). As a major interpreter of Mauss and Leroi-Gourhan for contemporary interests, Lemonnier (1989, 1992a) describes technical activity as the interplay of five heuristically separated elements: matter, energy, objects, gestures in sequence, and knowledge. Clearly, the first three elements share much in common with classic neo-evolutionary theory, wherein a cultural system is only considered as adaptive as its ability to transform matter and energy into efficient and viable cultural (survival) behaviors. The latter two elements of a technical activity, gestures in sequence and technical knowledge, derive from Mauss's concept of *enchaînement organique* (discussed next).[1] Taken together, the study of these five elements is supposed to permit an understanding of the sequential physical actions and decision-making strategies by which matter was transformed into culture-bearing objects. However, this chapter argues that only if we explicitly ask about social agency and the context-specific nature of social relations of material production can we comprehend the anthropological dynamics of technology (Dobres and Hoffman 1994).

Once the technical sequences by which artifacts were fabricated, used, and repaired are identified, *chaîne opératoire* research is employed most often to infer something of the abstract cognitive processes and underlying normative logic systems structuring those acts (see especially Pelegrin et al. 1988; Perlès 1992; essays in Renfrew and Zubrow 1994). Without doubt, the idea that preservable traces of artifact manufacture can be analyzed (through studying its technical chain) to identify deep-seated mental maps has provided archaeology with a major theoretical and analytic advance by researchers on both sides of the Atlantic (as in Lechtman 1977, 1984).

Nonetheless, a major strand in the web of interwoven social and material "threads" constituting technology lies somewhere *between* the sequential physical gestures of material culture production and use and the abstract cognitive frameworks structuring those activities. This particular thread concerns technicians, technical artifice, and dynamic interpersonal social relations of production (Dobres 1995a, forthcoming; Dobres and Hoffman 1994). For all the current focus on the structural "grammars" and the symbolic dimensions of prehistoric technologies, rarely do we explicitly theorize about the dynamic interactive social relationships through which all this symbolic and cognitive object-making and use took place. Even with more conventional interests in the "organization of technology" (typically pursued through cost-benefit analysis), explanatory models are rarely explicit on the dynamics of self-interested technical agents, the artifice of technical acts, much less the particular sorts of interpersonal relationships implied (Pfaffenberger 1988; Winner 1986). Without explicit attention to the socially constituted nature of technical acts and technical choice (those, for example, through which tacit mental maps were followed so faithfully or by which the physical and mechanical properties of raw materials were discovered or circumvented), neither the material nor mental processes involved will be adequately understood. Another way to think of this is that without attention to the hows and whys of everyday technical agency and the social contexts of those activities, descriptions will be little more than static (albeit sequential) rather than dynamic accounts of anthropological processes played out in and on the ancient material world.

Mauss's Enchaînement Organique *and Technology: A Total Social Fact*

Leroi-Gourhan's concept of *chaîne opératoire* derives from the ideas of the French ethnologist and sociologist Marcel Mauss (1935). As Durkheim's student, Mauss was interested in how social collectives articulated and maintained their mutually shared beliefs and traditional ways of acting in everyday life. Insightfully, he realized that technical acts were an integral part of the everyday way in which cultural traditions were maintained and passed on. At the same time, Mauss was interested in the body of collective knowledge underlying technical acts and how technical *connaissance* and *savoir-faire* were reaffirmed through routine physical gestures. For Mauss, then, technical knowledge involved more than understanding the physical properties (and limits) of raw materials and the practical knowledge enabling artifact production and use. It was also important to understand how technical *savoir-faire* was passed from generation to generation, embedded with value and significance, and reaffirmed through systems of kinship and apprenticeship. For Mauss, technology was dynamic and social to the core.

As an ethnologist, Mauss was especially interested in the *enchaînement organique* by which natural resources were sequentially transformed into useful cultural objects through bodily gestures practiced in socially constituted milieus. In keeping with Durkheim's lifelong interest in social solidarity, Mauss (1924) believed it necessary to move upward in analytic scale, from an understanding of the individual as *homo faber,* to that of the social collective and the enframing traditions within which individual techniques were practiced. He saw bodily techniques as integral to the everyday reproduction of society; thus he conceptualized technique as firmly embedded in normative cultural tradition (1935). Much like Heidegger (1977), Mauss argued the importance of understanding technical acts as they "unfolded," as they were in the process-of-becoming, for that was a key locus for the simultaneous production and reiteration of cultural meaning and practical action. This physical process of artifact-becoming-in-a-social-milieu is not only central to Mauss's concept of *enchaînement organique.* It is a physical link between artifactual and cultural reproduction that can be of special import to archaeologists. In sum, *chaînes opératoires* are of a decidedly social, collective, and material nature and show that technologies link together social, biological, gestural, and material transformative processes (Schlanger 1990).

Leroi-Gourhan: Methodologist and Evolutionist

It was André Leroi-Gourhan who provided these ideas with much-needed analytic rigor and introduced them into archaeology (Lemonnier 1992b). But in so doing, he also narrowed Mauss's total social fact to its more tangible side. As a student of Mauss, Leroi-Gourhan well understood that gestural sequences were rooted in ethnic traditions and entrenched in communal memory (Leroi-Gourhan 1964a:66). But as an archaeologist, he intellectualized technology at two vastly dissimilar scales: empirical traces of individual technical gestures, and macroevolutionary processes (Lemonnier 1992b:15–16; Pelegrin et al. 1988).

In his more philosophical reflections, Leroi-Gourhan explored the relationship between the somatics (or biology) of technical gestures and the technical act. In particular, he focused on the integration of bodily technogestures with the physicality of objects themselves (especially, 1943, 1945, 1964a, 1964b). But rather than explore how body and object integrated into an inseparable and total *social* phenomenon at the scale of everyday practice (as did his mentor), he concentrated on their evolutionary implications. For Leroi-Gourhan, the first meaningful act of distance and separation was between humans and their (material) technology, for example, in the evolutionary shift from the use of the hand *as* a tool to the hand *holding* a tool (Edmonds 1990:67; see more generally Ingold, in the Foreword to this volume). In contrast, Mauss focused on the ways *chaînes opératoires* were acquired through connections to the social body, through both direct and tacit education, *savoir-faire,* routinization, and even through self-awareness on the part of the working technician (see also Dobres forthcoming; Heidegger 1977; Ingold 1990, 1993a, 1993b).

Nonetheless, Leroi-Gourhan's introduction of *chaîne opératoire* into archaeological research began a critical and important shift away from the study of artifact morphology, typology, and function and toward an interest in the dynamic "life histories" of artifacts (see also Kopytoff 1986:84; Schiffer 1975, 1992b).

ON THE SOCIAL AGENCY OF *CHAÎNES* AND LINKS

Technologies are dynamic acts of social and material transformation: they serve as media through which social relations and world views are expressed and mediated; they materialize and make concrete people's attitudes about the right (and wrong) ways to make and use things; and, technologies take shape as they "take on" meanings and values by virtue of how technicians engage with each other while taking care of business (see also Pfaffenberger chapter 7 in this volume). In my desire to infuse the archaeological use of *chaîne opératoire* with a more dynamic human face, I find it especially useful, indeed necessary, to bridge the artificial barriers separating the many disciplines interested in technology from one another and from archaeology (Dobres and Hoffman 1994; Dobres forthcoming). Thus, I am in sympathy with philosophers, historians, sociologists, and sociocultural anthropologists who together define technologies as integrated *webs* weaving skill, knowledge, dexterity, values, functional needs and goals, attitudes, traditions, power relations, material constraints, and end-products together with the agency, artifice, and social relations of technicians.

For archaeologists, it is especially important to explore the parallel between technologies as acts of material transformation and technologies as acts of social transformation. As both Marx and Engels (1970:42–45) and Childe (1936) understood, through acts of material transformation people effect their own social transformation. And though this dialectic has long been recognized on a societal (evolutionary) level, theories of ancient technology have been relatively silent on how individual social identities were defined, expressed, and negotiated—that is, transformed—during technical acts. It is on this topic that *chaîne opératoire* can be of special methodological *and* interpretive value.

Contemporary theories of social agency can illuminate some of the above concerns by helping archaeologists understand how ancient technicians might have tried to situate themselves in relation to others while engaged in everyday technical pursuits (Dobres and Hoffman 1994; see also Cross 1990; Dobres 1995a, 1995b). In general, theories of social agency make sense of cultural practice as routinized and habitual actions in which women and men, children and elders engage on a day-to-day basis. As such, mundane and habitual social interaction lies at the heart of cultural practice and culture change. Modeling daily life at the microscale, therefore, becomes necessary to understanding more macroscale cultural and transformative processes (Marquardt 1992; see example in Roux and Matarasso chapter 3 in this volume). In the course of daily interaction, individuals express various kinds of interests, both personal and collective. In such arenas, the dynamics of interpersonal social relations inevitably lead to tensions and conflicts that require mediation and resolution. Nowhere is this *processual* dynamic more evident than in the daily practice of habitual technical activities.

For example, many ethnographic studies have shown how the agency of gender is inscribed onto the material world of resources and power, thereby affording certain individuals control of the objects produced, control of the technologies and technicians involved, control of the value systems that regulate the status of gendered technicians, and control of both esoteric and practical knowledge (see outstanding examples in Herbert 1993; McGaw 1996; MacKenzie 1991; Schwartz Cowan 1979). Therefore technologies are, at one and the same time, arenas in which agents construct social identities and forge power relations while also producing and using utilitarian objects for practical ends.

The conceptual framework of *chaîne opératoire* can link together the tangible and intangible dimensions of technological practice by regarding techniques as gestures undertaken in the "public" domain. At the same time, the analytic methods of *chaîne opératoire* research make it possible to link the archaeological record, comprised of static yet tangible remains of ancient technogestures, to the dynamic social milieus in which they were practiced. *Chaîne opératoire* research provides an excellent starting point for establishing such links, because it is specifically designed to identify and describe the material sequence(s) of gestural acts through which natural resources were modified (and remodified) into culturally useful objects. The idea in this chapter is that *while* undertaking productive activities, individuals create and localize personal and group identities, making statements about themselves that are "read" by others with whom they are interacting. Technical acts can thus be treated as a medium for defining, negotiating, and expressing personhood.

As Childs' ethnographic study in chapter 2 illustrates, social identity, status, power, privilege, and access to important cultural objects are inseparably interwoven facets of a single complex dynamic. A Toro man cannot "buy" a wife until he has access not only to the means and forces of iron production, but also to the esoteric and practical knowledge and skills necessary to become a viable iron maker. Only when these factors come together in specific times and places in the life histories of individual men can they acquire wives, a living, and social status. Can we possibly understand the complexity of Toro iron working without making these social dynamics and world views an integral part of materials analysis? Archaeologists need not restrict themselves to the observable present when asking about such processes. As an analytic method, *chaîne opératoire* provides the means for inquiring into such dynamics in prehistory.

Chaîne opératoire is explicit with respect to tracking the material and functional life history of artifacts. But as a corollary, *transforming natural resources into cultural products engenders the life history of technicians* (Dobres 1995b, forthcoming; Heidegger 1977). For example, as but one notable social identity cross-cut by many others, gender is not an immutable social category of person. What makes gender a cultural dynamic—that is, a process not an entity—is that it is *processed* throughout the life cycle (Moore 1986; Wolf 1974). Over the course of their lives, people move in and through salient cultural categories that conflate age, sex, and sexuality in complex ways: newborn, virgin girl, circumcised boy, pregnant wife, husband, skilled craftsman, father, mother-in-law, grandfather, divorcee, widow, and so forth. Gender, especially when conflated with age, has a sequential dimension not unlike a *chaîne opératoire.* Each gendered persona can, and typically does, confer new and different statuses on an individual. More to the point, gendered identities are processed, marked, and negotiated during one's life through the particular way individuals engage with one another while pursuing technical activities. Gender is, therefore, "manufactured" (de Lauretis 1987:9) in ways that resonate with the manufacture of material culture (for another example, see Lechtman's Afterword to this volume).

Attention to the ways individuals negotiate cultural categories does not ignore the fact that there are proscribed normative behaviors communities expect individuals to follow. Nor does it have to mean that social identities are something people consciously negotiate with every breath they take. Here is where Bourdieu's (1977) concepts of *habitus* and routinization are useful. They suggest how the seemingly inconsequential, mundane, and everyday acts of producing and using artifacts serve to habituate individuals to the codified social categories on which expressions of personhood may rest. It is in balancing expressions of self-interest against cultural norms and expectations that the dynamics of social relations of production become especially salient (see Hoffman chapter 5 in this volume).

THE DIALECTICS OF GENDER AND TECHNOLOGY

Once a conceptual parallel is established between making gender and making material culture, it becomes clear just how intertwined they are in everyday life. In particular, as McGaw (1996) and Pacey (1993) show, even what is defined a priori as "technology" stems in large part from gender ideologies and value systems, and that the two are inseparable:

> "Technology," like "economics," is a term conventionally defined by men to indicate a range of activities in which they happen to be interested. . . . Nearly all women's work, indeed, falls within the usual definition of technology. What excludes it from recognition is not only the simplicity of the equipment used, but the fact that it implies a different concept of what technology is about. Construction and the conquest of nature are not glorified, and there is little to notice in the way of technological virtuosity. (Pacey 1993:104)

There are few jobs that are not at least ideologically associated with particular genders, that do not differentially reward the players, and that do not also conjure up gender idioms. For example, in Western industrial societies men build and *mount* bridges, *erect* skyscrapers, and *man*-ipulate natural resources into cultural objects—all of which are unequivocal technical acts that simultaneously serve as powerful idioms of masculinity. In contrast, women's stereotypic share in codified divisions of labor turns on seemingly natural activities: having babies or nursing the young, sick, and elderly—none of which are typically thought of as technical per se, but which define femininity nonetheless.[2]

Significantly, until the advent of male-controlled Western reproductive technologies (in the sense of hard medical implements such as the speculum and, now, in-vitro fertilization), women's reproductive labor was thought of as natural and biological, but not necessarily technical. Yet even so-called natural childbirth has technical dimensions: to deliver a child safely and properly from the mother's womb requires a *chaîne* of biogestural techniques that vary cross-culturally. Such techniques combine esoteric and practical knowledge with practical skill and are supported by a particular configuration of social relationships (be they with midwives, nurses, doctors, nutritionists, or shamans). And as with any technology, the end-product is not limited to the production of a healthy baby; it includes new social positions for the mother and father, as well as the social and material power both gain within their community and from members of their immediate family (for examples, see Wolf 1974; also Childs chapter 2 in this volume). Archaeological theory rests on the largely implicit premise that prehistoric lithic and ceramic production, big game hunting, and architecture all had their technical side. This "hard" concept of technology also suggests that less material and seemingly natural activities (such as childbirth and child rearing) were not properly technical. It is clear, here, that the primacy of the hard and the utilitarian conflates with contemporary gender value systems to define what counts as a prehistoric technology (Conkey and Spector 1984; Gero 1991; Lechtman 1993). This conflation is a prime example of what Whitehead (1927:73–86) called the "fallacy of misplaced concreteness."

Gender and Technology in the Communal Mode of Production

In the communal mode of production (CMP), the social organization of technological *production* (especially in terms of subsistence) involves material, economic, and political divisions of labor in which women and men differentially participate.[3] At the same time, *reproduction* in such societies involves divisions of labor that are said to derive from "natural" differences between the sexes, and is typically defined as activities reproducing labor itself (that is, biological reproduction) and related "domestic" activities (classic statement in Marx and Engels 1970:52). This distinction between material production and social reproduction, and between so-called political/public and domestic/private spheres leads to another problematic corollary: the kinds of work in which women characteristically engage, such as child rearing and maintenance of the domestic sphere, are neither economic, political, nor all that technical (Hartmann 1987; Pateman 1987; Rosaldo 1980).

Because women's contribution to reproduction in the CMP has not been seen as properly economic, political, or technical, many of their everyday strategies for building networks of influence, prestige, and status have not attracted the same degree of academic interest as the "hard" techniques men practice (Ortner and Whitehead 1981; Rogers 1975). There is one important exception here: where women have been found to contribute significantly to *subsistence,* theorists have noted well that such activities serve as techniques of identity, status, and power. For example, in many post-colonial hunter-gatherer and foraging societies in South Africa and Australia, women have been found to contribute upward of 80 percent of the group's daily nutritional needs through plant gathering and more reliable techniques for hunting and trapping small animals (Hawkes et al. 1989; Lee 1979:253–272, table 9.3). And indeed, in many cases where women contribute the lion's share of daily foodstuffs, they often enjoy an important degree of autonomy and independence that women in other modes of production do not (Lee 1982; Sanday 1981:133–134; but see Bloch 1975 for a sobering counterexample).

RETHINKING TECHNOLOGY: ENGENDERING THE *CHAÎNE OPÉRATOIRE*

Because the social agency of identity is implicated in people's lifelong engagement with their material world, the variety of technical activities individuals undertake should shed light on the sequential nature of their intertwined social and material lives. For archaeologists, making a conceptual link between the two sequences—one social, the other material—sets the stage for reasonable inferences about the production (and use) of social identities through material production (and use) activities. However, as we intertwine social and material life histories in our technical study of artifacts, we must also broaden what we mean by technology (Dobres and Hoffman 1994; Ingold 1988, 1990; Layton 1974; see chapter 1 in this volume).

For example, in suggesting that archaeologists study a broader set of factors impinging on the structure of hunting activities, as but one "properly" prehistoric technology, Gifford-Gonzalez (1993:190) cogently argues the "need to consider the imperatives of the household in driving field processing decisions." Another way to think of this, as feminist anthropologists have long argued, is that a separation of the "domestic" female sphere (thought to involve food processing, cooking, and care-taking) from the supposedly more "technical" and male-oriented world (of hunting) is a *construct*—a typological distinction arbitrarily separating reproduction from production (Rosaldo 1980). As Gifford-Gonzalez (1993:187–188) puts it, "the lack of attention to cooking and culinary end-products in zooarchaeology is, I believe, attributable to unconscious androcentric bias within the field. . . . This view favors hunting—especially male pursuit, dispatch, and butchery of prey—over just about any other activity involving animals." Following the logic of *chaîne opératoire,* ancient hunting techniques were inseparable from the antecedent value systems and material context(s) structuring them; thus the before, during, and after of "the hunt" was necessarily linked to concerns extending beyond physically killing game (see also Ridington chapter 8 in this volume for an extended discussion of contemporary Athapaskan hunting technologies).

In a study of the contemporary Iñupiat (Inuit), Bodenhorn (1990:55) argues that "hunting cannot be reduced to the catching and slaughtering of animals, but rather includes a whole set of activities, both technical and symbolic, in which the interpendence [*sic*] of women and men is fundamental." Though whale hunting is described in the traditional anthropological literature as an exclusively male domain (because only men go to sea and actually spear whales), from an emic point of view Iñupiat women are directly implicated in and responsible for whale hunts. To ensure men's success on the open sea, the captain's wife must carefully comport herself in proscribed ways around their house; after all, "the whale comes to the whaling captain's wife" (Bodenhorn 1990:61). As well, through her technical skill with an *ulu* (knife) she is charged with "calling up" the whale to be killed by men (Cassell 1988). It is thus the wife's job to attract the whale and, in Iñupiat terms, "to hunt" (Bodenhorn 1993:191). Ask a whaling captain and he will tell you: "I'm not the great hunter, my wife is" (Bodenhorn 1990).

As well, among the Chipewya, "the hunt" encompasses cosmological beliefs, gender ideologies, and gendered (not sexual) divisions of labor (Brumbach and Jarvenpa 1997; Sharp 1991:190). To the Chipewya, what is important about a hunt is not just the animal killed, but the tasks women perform to process and distribute meat and skins afterward. According to Chipewyan informants, it is the transformation of "raw" meat into cultural products (such as food and clothing) that is valued and gives value to hunting: acquisition techniques and the immediate end-product (meat) are of lesser importance (Sharp 1991). This example recalls Heidegger's emphasis on the "becoming" and processual nature of technology. It also shows why we need to broaden the current Western view of technology beyond practical concerns, efficiency, and material matters. By stressing the overlapping social and material acts of technique and technology, we are situated to put a human face to technological practices, past and present.

POLITICS, IDENTITY, AND TECHNOLOGY IN THE COMMUNAL MODE OF PRODUCTION

A key social dynamic that irrevocably links together material and social transformations is the politically charged nature of technological systems, technical acts, and end-products. Among other things, the politics of technology concern how the social labor of production is organized; who has access to the means, forces, and relations of production; what socioeconomic status(es) are at stake; and who is affected by their implementation (summarizing Bijker et al. 1987; Law 1991; Winner 1986). Studies of contemporary (industrial) technologies offer especially striking examples of technopolitics that link production, power, and personhood, but there is no basis for assuming a priori that such dynamics were not intertwined in the past as well (Bender 1989; Conkey 1991; Dobres 1995a; Gero 1991; Hayden 1994).

Knowledge and Enskilment: A Political Basis for Status

As a general rule, individuals in communally organized societies often gain status, prestige, influence, and power by demonstrating highly valued qualities for all to see. The possibility of a personal basis for position is, of course, offset by the proscribed roles individuals are expected to fill by virtue of their age, gender, kin and tribal affiliations, and so forth. Although fundamental social categories initially define one's identity(ies) and occupations, it is not through them alone that people find their place in the social collective. Such first-order identities are built upon and transformed by virtue of personality, level of skill and talents, personal history, reputation, and the possession and display of knowledge through techno-gestures. All of these can become the basis of social and material power (Keene 1991; examples in Ingold 1993c; Ridington 1988; Saitta and Keene 1990:205). Thus, in both tangible and intangible ways, technological practice is directly implicated in expressing political identity in egalitarian communities.

Knowledge, especially, figures prominently as a cross-cultural basis for leadership in nonhierarchical societies (Ingold 1993a, 1993b; Keene 1991; also Ridington chapter 8 in this volume), and it is worth reiterating that technology is the juncture of knowledgeable practice and practical knowledge (Ingold 1990; Schiffer and Skibo 1987). The *concrétisation* of ideas, values, and knowledge in technical practice and end-products provides an outward expression of the metaphors and world views central to, but not necessarily shared equally by, all members of a community (Childe 1956; Lechtman 1984; MacKenzie 1991; Simondon 1958). For example, one of the salient characteristics of most Canadian subarctic egalitarian societies (among them, Cree, Dene, Blackfoot, and Ojibwa) is that social status and the acknowledged capacity and right to leadership roles, however temporary and context-specific, are achieved through participation in vision quests and other forms of knowledge acquisition (Ridington 1988).

Moreover, contrary to both Rousseau's romanticized nineteenth-century view of primitive communism and Sahlins' (1972) neo-evolutionary concept of the original affluent society, people communally organized do *not* always work smoothly toward agreed-upon ends, even in egalitarian societies.[4] In particular, social tensions and acts of contestation during everyday routines of material production and use can become a means for expressing power and influence (Conkey and Gero 1991; Dobres 1995a). In a study of the organizational and power dynamics of production on the Israeli *kibbutz,* Keene (1991:377) shows that

> within the communal mode, the social group as a whole—the commune—serves as the basis for all productive activity. Access to necessary factors of production is guaranteed to all members, and all members participate in determining the division between necessary and surplus labor. *This pattern still leaves room for internal variation and does not necessarily demand material equality or equal access to the means of production.* (Emphasis added)

Anthropologists have paid considerable attention to the ways in which men in nonhierarchical societies develop social and material privilege, for example, through their skills as hunters (Hawkes 1993), through their exchange of women (Meillassoux 1981), and even through the production and subsequent hoarding of technical knowledge and skill (discussed next). Following Ingold (1993c), I suggest that the display and manipulation of cultural metaphors or practical knowledge signified outwardly in the performance of particular gestural techniques are also powerful "mechanisms" for negotiating social identity and status. Elsewhere in this volume, Childs (chapter 2), Hoffman (chapter 5), and Pfaffenberger (chapter 7) show how creating, possessing, and displaying technical knowledge and skill over the course of one's life can translate into social power and status. In communally organized societies, then, one need not have physical control "over" material techniques and products to exercise authority: differential access to technoscientific knowledge (Schiffer and Skibo 1987), enskilment (Pálsson 1994), and technical virtuosity (Moore 1981; Root 1983) will do as well.

THE POLITICS OF SOCIAL AGENCY IN PREHISTORIC TECHNOLOGY: TWO ARCHAEOLOGICAL EXAMPLES

My comparative study of bone and antler technology at eight broadly contemporaneous Late Magdalenian sites in the French Midi-Pyrénées (ca. 14,000–11,000 years B.P.) identified extraordinarily variable sequential operations of artifact production, use, and repair around the region. Site-by-site, the actual choice of technical sequences used to manufacture, use, and repair harpoons, spear points, needles, and the like was structured *neither* by artifact physics *nor* by functional necessity. The guidelines for making, using, and repairing objects of the same raw material and even those of similar function were not based on objective conditions. Even among similar classes of artifacts retrieved from similar "site types"—such as needles at base camps and harpoons at upland seasonal ibex hunting sites—on-the-ground technical practices varied along almost every measurable attribute (Dobres 1996). Across the region there was no one "best" (or singularly "Magdalenian") way to make, use, decorate, or repair harpoons, points, and needles. Technical decisions were site-specific.

This evidence contradicts the notion that Late Magdalenian hunter-gatherers went about their daily lives faithfully following normative technocognitive maps maintained through a conservative cultural work ethic. Although they may have carried in their heads a roadmap of techniques to get a job done, *in practice* their strategies varied considerably. What, then, can account for these observable patterns?

When technical agents work in communal contexts, they are at least tacitly aware of each other's bodies and actions (Graburn 1976). And as Mauss recognized, when people are so engaged, even subtle bodily movements become a communicative medium. How individuals comport themselves while undertaking everyday technical activities, then, becomes a form of silent social discourse. "The very practice of a technique is itself a statement about identity: there can be no separation of communicative from technical behavior" (Ingold 1993b:438). People express interests of many kinds while making and using material culture, and as a general rule, there is a cacophony of messages communicated through technical body language. Fortunately, ethnoarchaeological and replicative experiments on the spatial patterning of material *chaînes opératoires* allow archaeologists to delimit with some precision the different contexts in which ancient artifacts were fabricated, used, and repaired on a site-specific basis. Because they were so often undertaken in what I like to call public contexts, where a variety of productive activities were going on simultaneously, ancient technical gestures were surely a total social fact. Intended or otherwise, technogestures so contextualized served as silent codes of discourse localizing identities of various sorts.

The gestural dimensions of Late Magdalenian organic technology can be understood as a sort of body language acted out in differently constituted contexts of social and material interaction. The technical evidence I have identified at individual sites in the Haute-Garonne and Ariège, specifically, attests to a significant degree of choice in how Magdalenian *chaînes* were actually practiced. It further points to the likelihood that in the day-to-day pursuit of such activities personhood was mediated through material means. For example: at La Vache, the highly variable patterning of harpoon barb construction suggests that technicians had individualized strategies for making them either longer, sharper, more curved, or thinner than those of their neighbors; at Les Eglises and Mas d'Azil, only a few bone needles show the successful completion of a highly skilled technique for piercing them with perfectly round eyes, while others betray an obvious lack of competence that those "in the know" could not have failed to observe; also at La Vache, only a few individuals (perhaps one) whittled an extra set of depressions on the base of their spear points, and through this, demonstrated materially (yet without recourse to overt self-aggrandizement) a special knowledge of hafting tricks learned over time. These subtle empirical variations in technical, functional, and morphological attributes imply an individualized level of *enchaînement organique* that would have been both visible and meaningful to one's neighbors, for as Graburn (1976:21) notes: "In small-scale societies where everything is everybody's business, there is little anonymity, and most people would know the details of style, the aesthetic choices, and even the tools of their contemporaries." It seems likely that the sorts of observable differences in Late Magdalenian technogestural strategies practiced on the ground contributed to localizing and materializing the identities and statuses of the technical agents performing them.

In the Paris Basin (also during the Late Magdalenian), French researchers have similarly suggested that technical knowledge was differentially practiced and shared under a system of tutelage they call apprenticeship (Karlin and Pigeot 1989; Olive and Pigeot 1992; Olive et al. 1991; Pigeot 1987, 1990). *Chaîne opératoire* research has been able to identify variable skill levels in blade production at Etiolles (Essonne) and their differential spatial distribution around discretely separated hearths. These data have been used to suggest that a hierarchy of interpersonal relationships (within family units) created and defined differential access to the practical and manual skills required to fabricate expert blades. Through the extraordinarily rigorous spatial mapping of variable qualities of blade manufacture, retouch, and repair "life" sequences, these mundane and practical endeavors have been resituated in something of their original social milieu. In turn, they have permitted reasonable inferences about the technical agency of ancient technological practice.

DISCUSSION

In these two archaeological examples, the French analytic of *chaîne opératoire* research has been employed to identify a number of interrelated factors structuring Late Magdalenian organic and lithic technology. Keeping the focus on the contours of site-specific social interaction, both studies concentrate on empirical remains to infer the organizational dynamics accounting for them (see also Roux and Matarasso chapter 3 in this volume). While still in their infancy, these two attempts demonstrate that *chaîne opératoire* research can be helpful in elucidating the socially constituted and intertwined histories of artifacts and artifice, products and people, and material actions and sociotechnical agency. Analytically and conceptually, we are now in an excellent position to establish even stronger anthropological links between static artifact patterns, traces of technogestural sequences, and the dynamic social contexts through which they materialized.

The preceding discussion has placed inordinate emphasis on the word *process,* arguing that social identities, subsistence activities, and artifact production all come about through sequentially organized technical activities. Allusion to the process-ing, or transformation, of both personhood and products is intentional, because technology always involves the recursive making of culture, agents, and material culture. For me, this means technical research needs to concentrate more on the *interrelationship* of social and material factors that combine to produce end-products, be they artifacts, the hunt, or individuals. Agents who move in and through specific material activities are embedded in, and thus bring with them, an extensive array of concerns and interests. The implications of technology's links and *chaînes* are ironic: while the material nature of the archaeological record often makes the dialectic of artifact and artifice hard to remember, it is also this tangible body of evidence that serves as our link to the intangible processes once involved.

Technology is no less than a materially grounded arena in which social interaction and contestation mediate the "becoming" of social agents and their artifacts. Thought of this way, technology is the sequential intertwining of social and material experiences best captured in the word *artifice.* Whereas artifacts and gestures take on their social life during productive sequences, social life is made meaningful and tangible through peoples' sequential processing of their material world. For all their daily and seemingly mundane repetition, ancient techno-gestures expressed the artifice of personal and group interests, reaffirmed collective memory, and materialized cultural sensibilities. Especially for prehistory, the conceptual framework underlying the pragmatics of *chaîne opératoire* research can be a powerful interpretive tool for understanding how social identities and relationships were constructed and transformed in the technological arena.

NOTES

Aspects of this work were supported by the National Science Foundation, the Woodrow Wilson National Fellowship Foundation for Research in Women's Studies, a Chancellor's Dissertation-Year Fellowship from the University of California at Berkeley, and the University of California (Berkeley) Office for the History of Science and Technology. At sequential stages in transforming these ideas into a readable product, I have benefited from the advice of Laura Ahearn, Jason Bass, Meg Conkey, Julie Cormack, Chris Hoffman, Tim Ingold, Heather Lechtman, Kent Lightfoot, Valentine Roux, Ruth Tringham, and Alison Wylie.

1. Quite independent of the French school, Anglo-American archaeologists have developed the notion of prehistoric technoscience (especially, Kuhn 1995; Schiffer 1992a: 134–138; Schiffer and Skibo 1987; see also Keller and Dixon Keller 1996 on "stocks" of technical knowledge).
2. In a later section I consider other less-than-technical (but typically "female-linked") activities, such as cooking and its relationship to "the hunt."
3. The CMP is not synonymous with hunting-gathering-foraging societies (extended discussion in Dobres 1995b:119–158).
4. On informal expressions of power and influence in egalitarian societies, see Cobb (1993), Flanagan (1989), and Keenan (1981), among others.

REFERENCES CITED

Bender, B.

1989 Roots of Inequality. In *Domination and Resistance,* edited by D. Miller, M. Rowlands, and C. Tilley, pp. 83–95. Unwin Hyman, London.

Bijker, W. E., T. P. Hughes, and T. K. Pinch (editors)

1987 *The Social Construction of Technological Systems: New Directions in the Sociology and History of Technology.* MIT Press, Cambridge, Mass.

Bloch, M.

1975 Property and the End of Affinity. In *Marxist Analyses and Social Anthropology,* edited by M. Bloch, pp. 203–228. John Wiley and Sons, New York.

Bodenhorn, B.

1990 "I'm Not the Great Hunter, My Wife Is": Iñupiat and Anthropological Models of Gender. *Inuit Studies* 14(1–2):55–74.

1993 Gendered Spaces, Public Places: Public and Private Revisited on the North Slope of Alaska. In *Landscape Politics and Perspectives,* edited by B. Bender, pp. 169–204. Berg, Providence.

Bourdieu, P.

1977 *Outline of a Theory of Practice.* Translated by R. Nice. Cambridge University Press, Cambridge.

Brumbach, H. J., and R. Jarvenpa

1997 Ethnoarchaeology of Subsistence Space and Gender: A Subarctic Dene Case. *American Antiquity* 62(3):414–436.

Cassell, M. S.

1988 Farmers of the Northern Ice: Relations of Production in the Traditional North Alaskan Inupiat Whale Hunt. *Research in Economic Anthropology* 10:89–116.

Childe, V. G.

1936 *Man Makes Himself.* Watts, London.

1956 *Society and Knowledge.* Harper and Brothers, New York.

Cobb, C. R.

1993 Archaeological Approaches to the Political Economy of Nonstratified Societies. *Archaeological Method and Theory* 3:43–100.

Conkey, M. W.

1991 Contexts of Action, Contexts for Power: Material Culture and Gender in the Magdalenian. In *Engendering Archaeology: Women and Prehistory,* edited by J. M. Gero and M. W. Conkey, pp. 57–92. Blackwell, Oxford.

Conkey, M. W., and J. M. Gero

1991 Tension, Pluralities, and Engendering Archaeology: An Introduction. In *Engendering Archaeology: Women and Prehistory,* edited by J. M. Gero and M. W. Conkey, pp. 3–30. Blackwell, Oxford.

Conkey, M. W., and J. D. Spector

1984 Archaeology and the Study of Gender. *Advances in Archaeological Method and Theory* 7:1–38.

Cross, J. R.

1990 *Specialized Production in Non-Stratified Society: An Example from the Late Archaic in the Northeast.* Unpublished Ph.D. dissertation, Department of Anthropology, University of Massachusetts at Amherst.

de Lauretis, T. (editor)

1987 *Technologies of Gender: Essays on Theory, Film, and Fiction.* Indiana University Press, Bloomington.

Dobres, M-A.

1995a Gender and Prehistoric Technology: On the Social Agency of Technical Strategies. *World Archaeology* 27(1):25–49.

1995b *Gender in the Making: Late Magdalenian Social Relations of Production in the French Midi-Pyrénées.* Ph.D. dissertation, Department of Anthropology, University of California at Berkeley. University Microfilms, Ann Arbor.

1996 Variabilité des activités Magdaléniennes en Ariège et en Haute-Garonne, d'après les chaînes opératoires dans l'outillage osseux. *Bulletin de la Société Préhistorique Ariège-Pyrénées* 51:149–194.

Forthcoming *Technology and Social Agency: Outlining an Anthropological Framework for Archaeology.* Blackwell, Oxford.

Dobres, M-A., and C. R. Hoffman

1994 Social Agency and the Dynamics of Prehistoric Technology. *Journal of Archaeological Method and Theory* 1(3):211–258.

Edmonds, M.

1990 Description, Understanding, and the Chaîne Opératoire. *Archaeological Review from Cambridge* 9(1):55–70.

Flanagan, J. G.

1989 Hierarchy in Simple "Egalitarian" Societies. *Annual Review of Anthropology* 18: 245–266.

Gero, J. M.

1991 Genderlithics: Women's Roles in Stone Tool Production. In *Engendering Archaeology: Women and Prehistory,* edited by J. M. Gero and M. W. Conkey, pp. 163–193. Blackwell, Oxford.

Gifford-Gonzalez, D.

1993 Gaps in Ethnoarchaeological Analyses of Butchery: Is Gender an Issue? In *Bones to Behavior: Ethnoarchaeological and Experimental Contributions to the Interpretation of Faunal Remains,* edited by J. Hudson, pp. 181–199. Southern Illinois University Press, Carbondale.

Graburn, N. H. H.

1976 Introduction: Art of the Fourth World. In *Ethnic and Tourist Arts: Cultural Expressions from the Fourth World,* edited by N. H. H. Graburn, pp. 1–22. University of California Press, Berkeley.

Harding, S.

1987 Introduction: Is There a Feminist Method? In *Feminism and Methodology,* edited by S. Harding, pp. 1–14. Indiana University Press, Bloomington.

Hartmann, H. I.

1987 The Family as the Locus of Gender, Class, and Political Struggle: The Example of Housework. In *Feminism and Methodology,* edited by S. Harding, pp. 109–134. Indiana University Press, Bloomington.

Hawkes, K.

1993 Why Hunter-Gatherers Work. *Current Anthropology* 34(4):341–361.

Hawkes, K., J. F. O'Connell, and N. G. Blurton Jones

1989 Hardworking Hadza Grandmothers. In *Comparative Socioecology: The Behavioral Ecology of Humans and Other Mammals,* edited by V. Stande and R. A. Foley, pp. 341–366. Blackwell, Oxford.

Hayden, B.

1994 Competition, Labor, and Complex Hunter-Gatherers. In *Key Issues in Hunter-Gatherer Research,* edited by E. S. Burch Jr. and L. J. Ellanna, pp. 223–239. Berg, Providence.

Heidegger, M.

1977 *The Question Concerning Technology and Other Essays.* Translated by W. Lovitt. Garland, New York.

Herbert, E. W.

1993 *Iron, Gender, and Power: Rituals of Transformation in African Societies.* Indiana University Press, Bloomington.

Ingold, T.

1988 Tools, Minds, and Machines: An Excursion in the Philosophy of Technology. *Techniques et Culture* 12:151–176.

1990 Society, Nature, and the Concept of Technology. *Archaeological Review from Cambridge* 9(1):5–17.

1993a Technology, Language, and Intelligence: A Reconsideration of Basic Concepts. In *Tools, Language, and Cognition in Human Evolution,* edited by K. R. Gibson and T. Ingold, pp. 449–472. Cambridge University Press, Cambridge.

1993b Tool-Use, Sociality, and Intelligence. In *Tools, Language, and Cognition in Human Evolution,* edited by K. R. Gibson and T. Ingold, pp. 429–445. Cambridge University Press, Cambridge.

1993c The Reindeerman's Lasso. In *Technological Choices: Transformation in Material Cultures since the Neolithic,* edited by P. Lemonnier, pp. 108–125. Routledge, London.

Karlin, C., and N. Pigeot

1989 L'Apprentissage de la taille du silex. *Le Courrier du CNRS: Dossiers Scientifiques* 73:10–12.

Keenan, J.

1981 The Concept of the Mode of Production in Hunter-Gatherer Societies. In *The Anthropology of Pre-Capitalist Societies,* edited by J. S. Kahn and J. R. Llobera, pp. 2–21. Macmillan, London.

Keene, A. S.

1991 Cohesion and Contradiction in the Communal Mode of Production: The Lessons of the Kibbutz. In *Between Bands and States,* edited by S. A. Gregg, pp. 376–394. Southern Illinois University Press, Carbondale.

Keller, C. M., and J. Dixon Keller
1996 *Cognition and Tool Use: The Blacksmith at Work.* Cambridge University Press, New York.

Kopytoff, I.
1986 The Cultural Biography of Things: Commodization as a Process. In *The Social Life of Things,* edited by A. Appadurai, pp. 64–91. Cambridge University Press, New York.

Kuhn, S. L.
1995 *Mousterian Lithic Technology.* Princeton University Press, Princeton.

Law, J. (editor)
1991 *A Sociology of Monsters: Essays on Power, Technology, and Domination.* Routledge, New York.

Layton Jr., E. T.
1974 Technology as Knowledge. *Technology and Culture* 15:31–41.

Lechtman, H.
1977 Style in Technology: Some Early Thoughts. In *Material Culture: Styles, Organization, and Dynamics of Technology,* edited by H. Lechtman and R. S. Merrill, pp. 3–20. West Publishing Co., St. Paul, Minn.
1984 Andean Value Systems and the Development of Prehistoric Metallurgy. *Technology and Culture* 15(1):1–36.
1993 Technologies of Power: The Andean Case. In *Configurations of Power in Complex Societies,* edited by J. S. Henderson and P. J. Netherly, pp. 244–280. Cornell University Press, Ithaca, N.Y.

Lee, R. B.
1979 *The !Kung San: Men, Women, and Work in a Foraging Society.* Cambridge University Press, Cambridge.
1982 Politics Sexual and Non-Sexual in an Egalitarian Society. In *Politics and History in Band Societies,* edited by E. B. Leacock and R. B. Lee, pp. 37–59. Cambridge University Press, New York.

Lemonnier, P.
1989 Towards an Anthropology of Technology. *Man* 24:526–527.
1992a *Elements for an Anthropology of Technology.* Anthropological Papers No. 88. Museum of Anthropology, University of Michigan, Ann Arbor.
1992b Leroi-Gourhan: Ethnologue des techniques. *Les Nouvelles de l'Archéologie* 48–49:13–17.

Leroi-Gourhan, A.
1943 *Evolution et techniques: L'homme et la matière.* Albin Michel, Paris.
1945 *Evolution et techniques: Milieu et techniques.* Albin Michel, Paris.
1964a *Le geste et la parole II: La mémoire et les rythmes.* Albin Michel, Paris.
1964b *Le geste et la parole I: Technique et langage.* Albin Michel, Paris.

McGaw, J. A.

1996 Reconceiving Technology: Why Feminine Technologies Matter. In *Gender and Archaeology,* edited by R. P. Wright, pp. 52–75. University of Pennsylvania Press, Philadelphia.

MacKenzie, M. A.

1991 *Androgynous Objects: String Bags and Gender in Central New Guinea.* Harwood Academic Publishers, Chur, Switzerland.

Marquardt, W. H.

1992 Dialectical Archaeology. *Archaeological Method and Theory* 4:101–140.

Marx, K., and F. Engels

1970 *The German Ideology.* International, New York.

Mauss, M.

1924 *The Gift.* Translated by I. Cunnison (1967). W. W. Norton, New York.

1935 Les techniques du corps. In *Sociologie et psychologie, Parts II–VI.* Reprinted in *Sociologie et anthropologie,* pp. 365–386. Presses Universitaires de France, Paris, 1950; also in *Sociology and Psychology: Essays of Marcel Mauss,* Translated by B. Brewster, pp. 97–123. Routledge and Kegan Paul, London, 1979.

Meillassoux, C.

1981 *Maidens, Meals, and Money: Capitalism and Domestic Economy.* Cambridge University Press, Cambridge.

Moore, H. L.

1986 *Space, Text, and Gender: An Anthropological Study of the Marakwet of Kenya.* Cambridge University Press, Cambridge.

Moore, J. A.

1981 The Effects of Information Exchange Networks in Hunter-Gatherer Societies. In *Hunter-Gatherer Foraging Strategies,* edited by B. Winterhalder and E. A. Smith, pp. 194–217. University of Chicago Press, Chicago.

Olive, M., and N. Pigeot

1992 Les tailleurs de silex Magdaléniens d'Etiolles: Vers l'identification d'une organisation sociale complexe? In *La Pierre préhistorique,* edited by M. Menu and P. Walter, pp. 173–185. Laboratoire de Recherche des Musées de France, Paris.

Olive, M., N. Pigeot, and Y. Taborin

1991 *Il y a 13,000 ans à Etiolles.* Argenton-sur-Creuse, Paris.

Ortner, S. B., and H. Whitehead

1981 Introduction: Accounting for Sexual Meanings. In *Sexual Meanings: The Cultural Construction of Gender and Sexuality,* edited by S. B. Ortner and H. Whitehead, pp. 1–28. Cambridge University Press, New York.

Pacey, A.

1983 *The Culture of Technology.* MIT Press, Cambridge, Mass.

Pálsson, G.

1994 Enskilment at Sea. *Man* 29:901–927.

Pateman, C.

1987 Feminist Critiques of the Public/Private Dichotomy. In *Feminism and Equality,* edited by A. Phillips, pp. 103–126. Blackwell, Oxford.

Pelegrin, J., C. Karlin, and P. Bodu

1988 "Chaînes opératoires": Un outil pour le préhistorien. In *Téchnologie préhistorique,* edited by J. Tixier, pp. 55–62. Notes et Monographies Techniques No. 25. CNRS, Paris.

Perlès, C.

1992 In Search of Lithic Strategies: A Cognitive Approach to Prehistoric Chipped Stone Assemblages. In *Representations in Archaeology,* edited by J-C. Gardin and C. S. Peebles, pp. 223–250. Indiana University Press, Bloomington.

Pfaffenberger, B.

1988 Fetishized Objects and Humanized Nature: Towards an Anthropology of Technology. *Man* 23:236–252.

Pigeot, N.

1987 *Magdaléniens d'Etiolles. Débitage et organisation sociale.* XXV Supplément à *Gallia Préhistoire.* CNRS, Paris.

1990 Technical and Social Actors: Flintknapping Specialists at Magdalenian Etiolles. *Archaeological Review from Cambridge* 9(1):126–141.

Renfrew, C., and E. B. W. Zubrow (editors)

1994 *The Ancient Mind: Elements of Cognitive Archaeology.* Cambridge University Press, Cambridge.

Ridington, R.

1988 Knowledge, Power, and the Individual in Subarctic Hunting Societies. *American Anthropologist* 90:98–110.

Rogers, S. C.

1975 Female Forms of Power and the Myth of Male Dominance. *American Ethnologist* 2:727–757.

Root, D.

1983 Information Exchange and the Spatial Configuration of Egalitarian Societies. In *Archaeological Hammers and Theories,* edited by J. A. Moore and A. S. Keene, pp. 193–219. Academic Press, New York.

Rosaldo, M. Z.

1980 The Use and Abuse of Anthropology: Reflections on Feminism and Cross-Cultural Understanding. *Signs* 5(3):389–417.

Sahlins, M.

1972 *Stone Age Economics.* Aldine, Chicago.

Saitta, D. J., and A. S. Keene

1990 Politics and Surplus Flow in Prehistoric Communal Societies. In *The Evolution of Political Systems: Sociopolitics in Small-Scale Sedentary Societies,* edited by S. Upham, pp. 203–224. Cambridge University Press, Cambridge.

Sanday, P. R.

1981 *Female Power and Male Dominance: On the Origins of Sexual Inequality.* Cambridge University Press, New York.

Schiffer, M. B.

1975 Behavioral Chain Analysis: Activities, Organization, and the Use of Space. *Fieldiana* 65:103–174.

1992a A Framework for the Analysis of Activity Change. In *Technological Perspectives on Behavioral Change,* edited by M. B. Schiffer, pp. 77–93. University of Arizona Press, Tucson.

1992b *Technological Perspectives on Behavioral Change.* University of Arizona Press, Tucson.

Schiffer, M. B., and J. M. Skibo

1987 Theory and Experiment in the Study of Technological Change. *Current Anthropology* 28(5):595–622.

Schlanger, N.

1990 Techniques as Human Action. *Archaeological Review from Cambridge* 9(1):18–26.

Schwartz Cowan, R.

1979 From Virginia Dare to Virginia Slims: Womanhood and Technology in American Life. *Technology and Culture* 20:51–63.

Sharp, H. S.

1991 Dry Meat and Gender: The Absence of Chipewyan Ritual for the Regulation of Hunting and Animal Numbers. In *Hunters and Gatherers.* Vol. 2: *Property, Power, and Ideology,* edited by T. Ingold, D. Riches, and J. C. Woodburn, pp. 183–191. Berg, New York.

Simondon, G.

1958 *Du mode d'existence des objets techniques.* Aubier, Paris.

Whitehead, A. N.

1927 *Science and the Modern World.* Macmillan, New York.

Winner, L.

1986 *The Whale and the Reactor: A Search for the Limits in an Age of High Technology.* University of Chicago Press, Chicago.

Wolf, M.

1974 Chinese Women: Old Skills in a New Context. In *Women, Culture, and Society,* edited by M. Z. Rosaldo and L. Lamphere, pp. 157–172. Stanford University Press, Stanford, Calif.

7. Worlds in the Making: Technological Activities and the Construction of Intersubjective Meaning

BRYAN PFAFFENBERGER

IT IS A PITY THAT ANGLO-AMERICAN social and cultural anthropologists pay so little attention to technology, tending (as they do) to observe coffeehouses, marketplaces, brothels, and street corners, rather than mines, shops, and boatyards. In all of these places, the boatyards as much as the brothels, culture develops and grows (Hannerz 1993). Yet English-speaking social and cultural anthropologists have all but ignored technology for much of the past fifty years, leaving the study of technology and material culture to ethnographic museums (Pfaffenberger 1992; Sillitoe 1988).

One reason for social and cultural anthropology's inattention to technology is the baleful influence of one of the founding fathers of the discipline, Bronislaw Malinowski (1884–1942). Seeking to put anthropology on a more scientific footing, Malinowski rejected the "purely technological enthusiasms" (1935:I:460) of the collectors, nontheoretical ethnologists, travelers, catalogers, and wild-guess evolutionary theorizers who had previously dominated anthropology. Dismissing their speculative theories, Malinowski (1935:1:460) adopted what he termed an "intransigent position"—in retrospect, a very justifiable one—that the study of technology alone was "intellectually sterile." Parenthetically, Malinowski was to add that a study capable of situating technology in its social context would indeed prove interesting, but this caveat was little noted by his students—or his successors. Shrugging off ethnological museums and the study of technology as well, social anthropology in the United Kingdom and cultural anthropology in the United States proceeded to ignore technology for decades.

In this chapter, I try to recover a serious anthropological theory of non-Western technological activity in relation to the creation and reproduction of worlds of intersubjective meaning. And I emphasize the term "recover," for the strands of a very satisfying theory can actually be found in the sociocultural anthropology literature (if one looks diligently enough). Ironically, these strands lead back to none other than Malinowski himself, the *bête noir* of anthropological studies of technology. In his least read work, *Coral Gardens and Their Magic* (1935), Malinowski turns with genuine interest to the study of Trobriand material culture and technology and begins to develop what could well have become a profoundly important theory of technological activity and culture. As will be seen, though, Malinowski commits a surprisingly inept interpretive blunder, obvious even to a non-Oceanist such as myself, and in consequence fails to develop his material. Interestingly, the theory-that-might-have-been is subsequently developed by anthropologists, who, following Malinowski's lead, worked in Melanesia and tried to grapple with the problems Malinowski left unsolved. I would not claim that these scholars saw themselves to be developing an explicit theory of technological activity in relation to culture. That is my construction, not theirs. Still, this tradition amounts to an important theoretical contribution, one that turns out to parallel remarkably similar (but independent) advances in the sociology of technology (Bijker et al. 1987), ethnoarchaeology (Childs 1991; Hosler 1996), the French anthropology of techniques (Gosselain 1992; Lemonnier 1983, 1986, 1989; Leroi-Gourhan 1943, 1945; Mauss 1935; van der Leeuw 1993), and archaeology (Dobres 1995a, 1995b, 1996, chapter 6 in this volume; Dobres and Hoffman 1994; Lechtman 1977; Schiffer and Skibo 1997).

When these various strands of anthropological work are brought together, social and cultural anthropology can offer an interesting and useful theory of technological activity in relation to culture, one that directly addresses many of the issues with which archaeologists continue to grapple. My basic contention is this: in recounting this lost tradition of theory, the most powerful meanings are generated, not only by the symbolism that is encoded in artifacts, but even more forcefully by the sequential, socially embedded sociotechnical *activities* that produce these artifacts.

"YAMS HOUSED BETTER THAN HUMAN BEINGS"

In Malinowski's *Coral Gardens and Their Magic* (1935), technology makes an explicit appearance in chapter 8, where he turns to the "show storehouses" called *bwayma*, used to store the yams that are the principal currency by which Tobrianders gain prestige and launch enterprises. No visitor to the Trobriands could fail to be struck by these impressive buildings—log cabins built on stilts with lofty roofs shaped in the form of a Gothic arch. They are, Malinowski (1935:I:218) observes with perhaps a note of sarcasm, "higher and more imposing than the living-house; more lavishly decorated, more scrupulously kept in repair; surrounded with many more taboos and rules of conduct."

Malinowski begins his chapter on the *bwayma* with an apologetic sentence that shows his antipathy to the previous generation of material culture studies: "Personally I am interested in technology only in so far as it reveals the traditional ways and means by which knowledge and industry solve certain problems presented by a given culture" (1935:1:240). The problems to which Malinowski refers are not solely the practical problems of economy and ecology, although the raised structure is ideally designed both to prevent pilfering and to protect the yams from rodents and insects. Rather, the *bwayma* addresses the crucial storage and display needs created by a culture whose internal dynamics are solidly based on the accumulation, control, and distribution of yams.

To solve these problems, Trobrianders construct what Malinowski terms *show storehouses*. Distinct from smaller, utilitarian ones, the show storehouses are built with lofty, imposing dimensions, an elegant shape, and conspicuous position (1935:1:242). The result, noted by Malinowski, is that yams are better housed than human beings, a fact that tells a great deal about Trobriand culture—and serves, at least to some extent, to construct it. In their communicative role, the show storehouses enable the prominent display of one's accumulation; they serve as an "index and symbol of power" (1935:1:240). Here, Malinowski anticipates a theory of style that has dominated archaeology for decades, namely, the thesis that the "stylistic" aspect of artifacts (as opposed to their "functional" aspects) issues from the artifact's role in providing social information (Barton et al. 1994). One could argue here that the artifact's style stems from the need to provide a material expression of ideology (DeMarrais et al. 1996).

But Malinowski was to put his finger on a complementary and more advanced theoretical approach, in which the communicative functions of style can be seen as a *byproduct* of artifact design.

Malinowski's analysis of the *bwayma* incorporates those aspects of the *bwayma*'s style that communicate "power," not only because they provide a visible symbol, but also because these aspects serve to incorporate specific modalities of social relationships into the entire fabricative process. The entire artifact can be construed, Malinowski argues, as a mechanism for *enrolling* constituent social modabilities and *transforming* them so that they suit the purposes of a chief aspiring to power and prestige. In other words, as Malinowski puts it, the chief can make use of a *social template* that is *pre-energized,* by which he means that the various individuals and groups brought into the fabrication process are already committed to the basic belief structure that underlies the activity, know what is expected of them and accept their obligations, and possess at least some sort of conception of what they are supposed to do. When the chief puts the template into action, the participating individuals are obliged to carry out a step-by-step *sequence of fabricative and ritual activities,* which run "parallel" to each other in a "strictly determined" order (Malinowski 1935:1:470). In this sense, Malinowski anticipates the French anthropological concept of *chaîne opératoire* (Gosselain 1992; Leroi-Gourhan 1943, 1945; van der Leeuw 1993), but in a way that avoids the technicism of Leroi-Gourhan's formulation and highlights the imbrication of technical, political, and ritual activities (see Dobres chapter 6 in this volume).

A "PRE-ENERGIZED SOCIAL TEMPLATE"

Malinowski's description of the "pre-energized social template" and its embedding in a tightly connected series of technical and magical operations is, I believe, the most valuable contribution Malinowski makes in *Coral Gardens.* A show storehouse is not just built by anyone. It is built by a chief who wishes to improve his rank—and by extension, that of his lineage. To construct the *bwayma,* the upwardly mobile chief begins by summoning his male relatives-in-law, men who will subsequently fill the storehouse with the yams presented in the famous *urigubu* transactions detailed in Malinowski's more widely read books (such as *Argonauts of the Western Pacific*). (In these transactions, men are obliged to make annual yam payments to their sisters' husbands, thereby putting themselves somewhat at odds with their wives and children—but such are the soap operas of matrilineal societies.)

As Malinowski goes deeper in his account of the *bwayma,* it becomes increasingly evident that something more than a mere building is being constructed. Under construction (along with the building itself) are the social and political relationships on which the chief must rely if he is to make a bid for higher rank; he must transform his kinship relationships into a well-oiled, extensive machine capable of producing astonishing numbers of yams. Building the *bwayma* provides an opportunity to "trigger" this machine, as well as to test its condition, strength, and extent (Tambiah 1990:73). In addition, the tightly connected technical and ritual activities provide an opportunity to drive home the patterns of cooperation, authority, and deference that must exist if the chief hopes to achieve his aim.

"AN ANCHORED CANOE"

In turning to the *bwayma's* symbolic dimension, Malinowski provides a richly detailed account. But, as will be seen, his reductionist explanation of magic as a means of reducing anxiety derails his analysis.

What does the *bwayma* symbolize? In brief, an anchored canoe. Malinowski discovers this by exploring the meanings related to the constituent parts of the architectural structure. The transverse logs that provide the *bwayma*'s foundation are called *kaylagim,* recalling the transverse boards (*lagim*) that enclose the well of a canoe at both ends; thus the *byawma* is a canoe, a metaphor echoed prominently in magical formulas. But this is not to be a light canoe that could skim over the surface of the sea; it is to be heavy, laden, dark, and anchored, as will be seen in the ceremonies that occur when the *bwayma* is filled.

When construction is finished, the lattice-like *bwayma* structure, transparent through the large openings between the logs, conveys an impression of airiness and lightness—an impression that, while functional for economic and ecological purposes, holds out compelling dramatic prospects. For filling the *bwayma* will turn this light, transparent structure into one that is dark, heavy, and anchored, all terms of approbation in the Trobriand scheme of things. Packed to the rafters with yams, the structure becomes opaque and dark, giving the impression of solidity and weight.

The rite that precedes the filling of the *bwayma,* called *vilamalia,* is a ritual of plenty and prosperity that echoes the theme of transformation. The *vilamalia* serves to "anchor the canoe": enumerating every component of the *bwayma,* the magician recites, "It shall be anchored," like immovable stone, like bedrock. It is transformed from something that is light and airy into something that is heavy and dark. For Trobrianders, heaviness and darkness connote prosperity and fertility, Malinowski explains.

But here Malinowski's interpretation begins to break down. Under the apparent influence of Victorian anthropological definitions of "sympathetic magic," Malinowski stresses the "toughness and tenacity" of the various substances that Trobrianders include in the *vilamalia* rites. He concludes that these rites are intended to ensure the fertility of the yam crop but notes with incredulity and amusement that the Trobrianders themselves emphatically disagreed. Bagido'u patiently explains to Malinowksi:

> Suppose the *vilamalia* were not made. Men and women would want to eat all the time, morning, noon and evening. Their bellies would grow big, they would swell—all the time they would want more and more food. I make the magic, the belly is satisfied, it is rounded up. A man takes half a tytu and leaves the other half. A woman cooks the food; she calls her husband and her children they do not come. They want to eat pig, they want to eat food from the bush, and the fruits of trees. *Kaulo* (yam-food) they do not want. The food in the *bwayma* rots . . . 'til the next harvest. Nothing is eaten.

Another villager echoes Bagido'u's statement: "When we do not make the magic of prosperity, the belly is like a very big hole—it constantly demands food. After we [perform the rites] the belly is already satisfied" (1935:1:227).

Where did Malinowski go wrong? Astonishingly, Malinowski never *explicitly* draws the connection between the *vilamalia*'s symbolism and the overriding *ethos* of this redistributive culture, in which giving yams—rather than retaining them for oneself—constitutes the foundation of the entire society. One possible reason: Malinowsi had not shaken himself free from the influence of the Victorian anthropologists, Tylor and Frazer, who saw the magico-religious dimensions of preindustrial technology as little more than a prescientific intellectual error. Malinowski (1954) goes on to develop his theory, easily refuted (Tambiah 1990), that the use of magic stems from the psychological anxiety of a primitive people faced with uncertain weather and the potential failure of a crop; it is for this reason that he ignores Bagido'u, and the other Trobrianders, and insists that the rites are about ensuring the fertility of the earth.

PARTICIPATING IN A FOUNDATIONAL MOMENT

If the statements of Malinowski's informant are taken at face value, constructing a *bwayma* is not so much about ensuring the fertility of the earth as it is about bringing about a transformation in people. It transforms them from gluttons who would eat every yam in sight into the yam refusers that the society must have if it wishes to remain viable. But how?

The theme of transformation appears throughout the sequence of activities that culminates in stuffing a *bwayma* to the rafters with yams. As has been seen, the building itself is symbolically an anchored canoe; the *vilamalia* rites repeat this theme by anchoring every conceivable component of the structure. But what is this anchored canoe? From Malinowski's own work, as well as that of subsequent ethnographers in the Trobriands and elsewhere in the Melanesian cultural area, it seems reasonable to suggest that it is an idealized representation of society itself. The theme of canoe-anchoring as a metaphor for the establishment of a prosperous, dignified society appears throughout Melanesian thought, and appears even in humor: a clan enjoys telling the hilarious story of its founder, a man so stupid that he did not realize that his canoe had run aground until neighbors pointed out that crops were growing in it (Damon 1989). In various guises throughout Melanesia, the anchoring of a canoe is a *foundational moment,* the key thing that must happen before the good life can be achieved.

With this point in mind, it is easier to understand why building the *bwayma* transforms people into yam-refusers. To build the *bwayma* is to participate actively in society's moment of birth, the mythological event of anchoring that gave rise to the clans and to everything that is considered of value to Trobrianders. To participate in all the various technical and ritual activities that culminate in the stuffed *bwayma,* is to walk through all the transformations—personal and social, as well as material—that must occur if Trobriand Islanders hope to ensure their society's viability. Extending Malinowski's valuable insights about the *bwayma,* then, it appears that building a show storehouse provides not only a means of calling forth and "triggering" indispensable *social* relations, but also indispensable *meanings.* The show storehouse's stylistic aspects and associated magical practices turn out to be part of an underlying and deeper social *process,* in which the total artifact—the building, the rites, and the ceremonies—constitutes an elegantly designed mechanism for triggering pre-energized social relationships, bringing them into play under the chief's supervision, and transforming the participants so that they fully understand what is required of them. Interpreted this way, the *bwayma's* communicative and symbolic roles are decidedly secondary to the role the total artifact plays in *enrolling* and *transforming* people.

STEPS TOWARD A THEORY

Although Malinowski did not fully exploit the evident connections between the *vilamalia* rituals and the reigning redistributive ethos of Trobriand Island society, his work suggests the beginnings of a theory that helps to explain why technological activities, far more than the symbolic messages embodied in the resulting artifacts, are powerful generators of intersubjective meaning.

Such a theory would stress Malinowski's concept of a "pre-energized social template" that draws upon existing social roles and obligations and enables people to solve a problem that their culture creates. When set into motion, this template marches the involved individuals through a conventionally fixed sequence of technical and ritual activities that culminate in the finished artifact. In terms of the symbolic aspects of the *bwayma,* the outcome of this process is not only the transformation of nature, so that it is more useful for human purposes, but also the transformation of people, so that they are better able to participate in their community (see similar argument in Dobres chapter 6 in this volume). This is achieved by linking the work processes as well as the generated artifact to an idealized representation of society, or more pointedly in some cases (as will be seen) to an idealized self. In this way the entire process amounts to *participating* in the society's foundational moment, namely, the creation of the type of *personhood* that the community deems essential to its viability. Represented throughout are "bad selves" (the type of person the community cannot tolerate) and "good selves" (the type of person the community *must* have). What drives all this home to the participants is the fact that it is experienced kinesthetically, by actually *performing* the meanings implicit in the activity's social template.

What remains to be seen here is whether such a theory can be fruitfully applied elsewhere. I explore this question briefly in two cases: canoe-making on the Melanesian island called Gawa, and coal mining in nineteenth-century Cornwall. The first case shows how a more recent anthropological interpretation of Melanesian technology avoided the mistakes Malinowski made, and the second highlights the potential cross-cultural utility of this theory.

FROM THE TROBRIANDS TO GAWA

Gawa, an island of some 400 souls at the time of anthropologist Munn's study, is famed for hundreds of miles in all directions for the beautiful seagoing canoes its inhabitants make. Gawans fabricate canoes by hollowing out logs, a laborious process that can take up to seven months to complete. But they do so only to part with the beautifully adorned vessels, for canoes constitute Gawa's chief contribution to the famous *kula* ring exchanges first described by Malinowski (1922): a "circle" of islands linked by the exchange of armshells in one direction and of necklaces in the other.

When a Gawan gives up a canoe to a *kula* trading partner on another island, he receives in return a trinket of personal adornment—the armshell that permits him to enter into the circular network of *kula* exchanges, to the glory of his clan and to the fame of Gawa. The point is not to acquire wealth, for *kula* items are, in the end, little more than trinkets. The point is to raise the rank of one's clan, a matter that can indeed carry with it significant consequences for the internal distribution of prestige on Gawa.

In 1974 Munn published a brilliant paper (albeit in an obscure journal) on the fabrication of Gawan canoes, in which she tried to show how this fabrication process helped create the world of intersubjective meanings in which Gawans live. What is fascinating about Munn's work, especially when considered against Malinowski's, is just how she avoided the traps into which Malinowski fell. What is more, to the extent that one can craft something approaching a theory out of the *bwayma,* it quite evidently applies with stunning precision to the making of canoes on Gawa.

A THEATER OF MORAL CONFLICT

To grasp the significance of Gawan canoe-making, it is important to realize, as Munn (1985:272) subsequently stressed, that this technical activity reveals to

Gawans *the type of moral conflict in which they are involved.* In this sense, Munn has chosen a starting point that is somewhat akin to Malinowski's. Recall that Malinowski was interested in technology only insofar as it could be shown to address "problems" that a culture poses. But "moral conflict" is a far richer concept than "problem" and leads Munn to recognize immediately the coherence of the meanings underlying technology in a way Malinowski never achieved.

As Munn depicts the fundamental moral conflict of Gawan social life, it is immediately clear that we are still in Melanesia; these are many of the same themes with which Trobriand Islanders are concerned, but Gawans—like other Melanesians (Damon 1989)—put their on spin on them. In the Gawan scheme of things, excessive eating makes the body heavy and slow and dull; stuffing oneself lies at the opposite extreme from the positive values of youthfulness, buoyant lightness, and feelings of joy (Munn 1986:76). In overeating, one takes on the heaviness of the soil, a quality that is positive in gardens but not in people. One becomes indolent; one's garden dies; one cannot give to others.

Restricting one's consumption is the necessary condition for creating a surplus, and ultimately creating Gawa's fame. As one Gawan put it, when someone eats a lot of food

> it makes his stomach swell; he does nothing but eat and lie down; but when we give food to someone else, when an overseas visitor eats pig, vegetable food, chews betel, then he will take away its noise, its fame. If we ourselves eat, there is no noise, no fame, it will disappear, [become] rubbish, it will default. If we give to visitors, they praise us, it's fine. If not there is no fame, Gawa would have no *kula* shells, no men of high standing, no *kula* fame. (Munn 1986:49)

FROM DARK TO LIGHT

How does fabricating a canoe bring about the needed transformation in people? Interestingly, Gawan canoe-making is very nearly the exact mirror image of *bwayma*-making. In creating the *bwayma,* Trobriand Islanders bring a light canoe *to* the island, make it heavy, and anchor it to the earth. In creating a canoe, Gawans take a heavy, dark log *from* the island, lighten it, and send it to sea. At every step of the canoe's fabrication, the various rites (not to mention the technical acts of construction) are intended to transform the canoe from a dark, heavy log into something light and airy, a vessel that can (as Gawans put it) "fly" across the surface of the water. To acquire the requisite propulsion, Gawans hollow out the log and also use magic, which Munn describes as a kind of "kinesis" that is "given by humans to wood."

At the core, the canoe's purpose is not so much to get from Point A to Point B but rather to create fame. This is accomplished not only by creating a beautiful boat, but also by creating beautiful people. The theme of personhood is impossible to miss, because the canoe is, symbolically, an idealized person. Parts of the canoe are playfully likened to parts of the human body; it has eyes, ears, shoulders, a chest, hair. And this is a beautiful, bright person, whose painted exterior "shines like lightning." In this guise, the canoe is like the mythical young man of magical powers who, emerging from the sea, shines forth as if he were reflecting a flash of lightning and, in so doing, captures the heart of any woman who chances to behold him in his apotheosis. Moreover, the canoe's gaily flowing pandanus streamers "call" to the *kula* partner, "turning his mind" so that it is partial to yielding up his armshells and engaging in exchange with Gawa; thus Gawa's fame "climbs" like blowing pandanus streamers. The canoe objectifies the subjective notion of the self that is essential for *kula* exchange; a man becomes "light" by rejecting self-indulgence, by giving to others, and finally by creating an adornment, only to part with it, but he receives an adornment in return, an armshell, which is testimony to his fame.

Gawan canoe-making may reverse the symbolism of the *bwayma,* but it accomplishes the same end. In making canoes, Gawans construct not only the canoe, but also themselves, by means of their *participation* in a "theater in which good selves and bad selves are equally represented" (Munn 1986:6). Once again, fabrication of a key material object—here, the canoe—amounts to a representation of society at its foundational moment. In both cases, the constructed object and its "becoming" symbolically represents the transformation that is deemed to occur; the *bwayma* stands for agricultural society that flourishes once the "canoe" is firmly anchored, while the Gawan canoe represents a person who becomes generous in the movement from dark to light. In light of this interpretation, it seems clear that the canoe's symbolic aspects are intended not so much to convey important social information, save to would-be *kula* trading partners on other islands, but rather to provide a field of kinaesthetic experience, in which people can experience the transformation of the canoe from dark to light, even as they experience this transformation in themselves.

"THE UNIFYING DIRECTIONALITY OF THE CANOE'S IRREVERSIBLE PATH"

In her analysis of canoe fabrication, Munn extends and develops Malinowski's notion of a "pre-energized social template" that can be brought into play—"triggered"—as needed. In Munn's view, canoe-making not only draws upon existing social and economic relationships, it also *modifies* them and repackages them in a way elegantly suited to the multiple objectives of the fabrication process. The chain of technical operations that creates the canoe involves a subtle manipulation of existing patterns of gift-giving, a manipulation that serves to "propel" the canoe off the island so that no one can claim further ownership of it.

The man who takes on the seemingly self-sacrificial task of fabricating the canoe begins by calling forth the labor of his matrilineal clanmates. Along with their spouses, they form the construction team (men build the canoe, while women cook for the workers). But this labor, Munn notes sardonically, must be "compensated"—nowhere else would men compensate women for cooking services—and so, on completion of the canoe, it is given away, perhaps to the builder's wife, and this act leads to a series of prestations and counterprestations in which pigs and taro pudding flow one way, and the canoe the other. The exchanges "launched" by the canoe serve to integrate internal marriage exchanges, transposing them on the "unifying directionality of the canoe's irreversible path" (as Munn pus it). In this way, Gawans construct a "new," synthetic framework of social relations, which cuts across the grain of the usual segmentary exchange patterns, while they construct the economic trajectory that will eventually "launch" the canoe off the island. The network of transformed gift-giving culminates in the gift of the canoe to a *kula* partner on another island, "fixing" the island's location in the interisland system of prestations and counterprestations that constitutes the *kula* ring.

THE TRIBUTE SYSTEM IN CORNISH MINING

If Melanesian ethnography could be construed to offer a theory of how technological activity generates cultural meanings, can such a theory be applied elsewhere—say, to nineteenth-century Cornish metal mining (Rule forthcoming)—and with fruitful results? Indeed, the theory works well, even in such a radically different context.

Reaching its peak in the mid–nineteenth century, Cornish miners developed a distinctive subculture characterized by extreme political apathy, disinterest in labor organization, and a highly emotional form of Methodism. In the chapels, "stout-hearted men from the mines come to prayer meetings" and are overcome; "they fall down on their knees and wrestle 'til they have found mercy . . . the round and piercing cries of the broken-hearted penitents drowned out the voice of prayer . . . the seats were covered in tears" (cited in Rule forthcoming). The apathetic, other-worldly culture of Cornish mining communities frustrated socialists and union organizers to no end (Rule 1992). Socialist intellectuals attributed the miners' political apathy to the stultifying influence of Methodism (the late E. P. Thompson was to characterize Cornish religiosity as "ritual psychic masturbation"). But it was not so much Methodism as the social relations and social experience of mining itself that seems to account for the extreme other-worldly religiosity and political apathy of Cornish miners. A key point of evidence: after the 1870s, even as Methodists were trying to make their denomination more respectable, ministers could not rein in the miners' emotional outbursts (Rule forthcoming). As the mining industry declined in subsequent years, Cornish miners emigrated to the United States and Australia, taking with them their culture of resignation, political apathy, and emotional religiosity.

In an attempt to explain the political apathy and emotional religiosity associated with Cornish mining, Rule (forthcoming) focuses on the actual physical dangers of coal mining, which appears to have been the most dangerous occupation of any pursued by significant numbers of workers in nineteenth-century Britain. Bereavement, widowhood, and orphanhood were common, so it is hardly surprising that Cornish miners would be especially receptive to a spiritual message of suffering, redemption, and the ineluctable working of Providential will. Without contesting this interpretation, it is useful to stress (as does Rule) how the specifics of the Cornish mining social template additionally helped drive home that they were living in a world of incomprehensible and uncontrollable chance, thus creating the fertile soil that Methodism was to till.

The social template of Cornish mining is called the *tribute system* (Lewis 1908; Rule forthcoming). Designed to ensure that labor and capital shared the risks of exploration, the tribute system compensated miners by providing small groups (often as small as two men and a boy) with a contract to explore a measured portion of the mine, called a *pitch*. The miners would share an agreed percentage (expressed as shillings per pound) of the price fetched by any discovered ore. Pitches were made available by auction, and contracts went to the groups with the lowest bids. Contracts ran for a fixed length of time, generally two months, and could not be renewed at the same rate.

Like the social templates discussed by Malinowski and Munn, the Cornish tribute system is founded in a culturally defined moral conflict: namely, the problem of capital versus labor as formulated by nineteenth-century economists:

> The system was acclaimed by those nurtured in the axioms of nineteenth-century political economy. It promised a high degree of task application because the miners were paid by results; and it associated labour with capital in the risks of enterprise. It naturally regulated the price of labour by supply and demand, for when work was scarce competition was fierce, and miners in effect bid down their own wages. Certainly contemporaries noted that involving a miner in

> periodic competition with his comrades was hardly propitious for collective action, and contributed to the weak development of trade unionism in the mines. (Rule forthcoming)

In short, the tribute system evolved in such a way that it served not only to produce a technological result (ore), but also to *discipline* workers, and to transform them into the type of individuals that were deemed to be necessary for society's viability. In effect, the tribute system constituted a representation of an idealized society, in which "bad selves" (wicked, impecunious miners given to agitation and sabotage) could be transformed into "good selves" (good, hard-working, miners who would share the risks of exploration). Once again, sociotechnical relations of production are not merely conventional means of performing a given task; what is more, they are a means of ensuring that people *experience* a transformation that is held to be vital to the society's overall viability. Like the social relations of Trobriand *bwayma* construction, the social template of Cornish Mining was "pre-energized" and "triggered" by drawing from conventional modes of social, economic, and legal relationships, including partnership, kinship, auctions, speculative investment, markets, and contracts; but it weaves them together into a unique and distinctive totality—an *artifice* (Dobres: chapter 6 in this volume), resulting in a sociotechnical phenomenon that actors themselves recognize to be distinctive.

And how does the tribute system transform miners? Cornish mining served to drive home a social experience well captured by the famous line from *Ecclesiastes:* "Time and chance happeneth to them all." Because of the inherent dangers of mining, this is something all miners experience to a degree; what is remarkable about the tribute system is the way this experience was driven home in every conceivable manner. For miners (as described by Rule forthcoming), the tribute system held out a chance of striking it rich—or, alternatively, of sinking deeply into poverty. In order to succeed, miners would have to discipline themselves and make rational choices, including choices of partners, the amount of capital to be invested in tools and equipment, and the amount to bid on a given pitch. At the end of the day, though, luck would determine whether a given pitch turned out to be better or worse than predicted, and the ore's selling price fluctuated wildly in a speculative market that was beyond anyone's grasp or control. Very rarely, a group would strike it rich, discovering a previously unguessed lode in a pitch acquired for a very low rate; when this happened, it served to inspire other groups to keep trying, even though such discoveries were rare. More commonly, a group that bid too high on a pitch that proved disappointing would find themselves and their families driven into deprivation. But miners did not blame the tribute system; instead, they attributed their fortune, good or bad, to the ineluctable workings of Providential will (Rule forthcoming). The miners' response was not to challenge the basic premises of the system, but rather to cast themselves to the earth before God, and beg His forgiveness and mercy. If the embodied intention of the tribute system was to produce miners who did not contest the economic arrangements of mining, this system succeeded very well indeed.

What is striking about this example is its minimization of the "Great Divide," that gulf presumed to separate the (industrial) capitalist mode of production from (preindustrial) precapitalist modes. At first blush, the Trobriand *bwayma* might seem to be an egalitarian affair, rooted in a community's unaffected self-expression of deep-seated cultural values, while the Cornish tribute system might seem to be an imposed system, cynically devised to discipline workers. What such a contrast hides is the very real kinship between them: both arise from attempts by social agents to draw from existing meanings and social relations, and weave them into a unique sociotechnical totality—an *artifice,* as Dobres (chapter 6 in this volume) adroitly terms it—that transforms those who experience it. In this sense the capitalists of Cornwall and the big men of Melanesia have something in common.

WORLDS IN THE MAKING

A lost tradition of anthropological theorizing about the role of technology in generating worlds of intersubjective meaning makes it possible to show that these meanings are generated not only by the communicative role of artifacts, but also—and perhaps much more powerfully—by the social experience of *making* them. According to this theory, in their material conditions technological activities embody a social template designed to solve a culturally formulated problem or moral conflict. The activities themselves are designed not only to accomplish their technological ends, but, what is more, to transform and discipline *people* so that they become the type of people the community thinks is required if it is to be viable. In this sense, the work process amounts to a kind of theater in which, as Munn puts it, good and bad selves are equally represented. The sociotechnical activities themselves amount to a comedy (Gawa) or tragedy (Cornwall) in which the represented selves are transformed, as they must be if the society is to be viable. In light of this theory, it is hardly surprising that so many human artifacts symbolically embody conceptions of personhood (David et al. 1988; example in Childs chapter 2 in this volume) or encode representations of social relationships (see Hoffman chapter 5 in this volume); the artifact "stands for" the transformed people and society, and in so doing objectifies the transformation even as it creates it and represents it. The meanings embodied in artifacts and their technical transformations, then, provide participants a way of reflecting to themselves (and kinaesthetically experiencing) the very meaning that they are actively constructing.

This theory implies that the relationship between culture and artifact-embedded meanings should be reexamined. It suggests that the symbolism so commonly inscribed in artifacts is not so much the *cause* of cultural meanings, as is so often supposed by studies that stress the communicative and informational roles of an artifact's style. Rather, its symbolism is the *consequence* of the *simultaneous* production of artifacts and meanings in patterned technological activity. This is theoretically consistent with the impressive evidence, summarized by Miller (1987), that artifacts are susceptible to differing interpretations by the various constituent groups of a society (also argued by Hoffman chapter 5 in this volume); women, for instance, may read artifact-embodied meanings differently from men (for a Melanesian example of contrasting male and female "readings" of war canoes, see Barlow and Lipset 1997). Any culture that relied on artifacts to communicate crucial meanings would quickly find that these meanings could spiral out of control. To make the meanings stick, people must be guided through patterned relationships in which they *experience* the desired meaning and no others. The theory further suggests that the meanings embodied in artifacts do not necessarily reflect beliefs that everyone in the producing culture accepts without question (Dobres chapter 6 in this volume). On the contrary, the expressed symbolism develops when agents weave existing meanings and social relationships into a unique sociotechnical totality in such a way that shared beliefs may be subtly modified, or combined in novel ways. Artifacts embody meanings created at a particular time and for a particular purpose by a particular technical collective, and they may very well exaggerate, misrepresent, or recombine in novel ways meanings that are actually shared by the society at large. In consequence, one cannot simply decode the meanings of an artifact and read backwards to shared cultural meanings (*contra* Prown 1982).

Above all, this approach suggests that when we engage in technological activities, we are constructing not only artifacts, but ourselves as well (Dobres 1995a). Whoever fully understands and is able to shape these activities possesses the power to determine what kind of self, and what kind of society, will be constructed.

NOTE

This essay was written with the assistance of a fellowship from the Swedish Collegium for Advanced Study in the Social Sciences, Uppsala, Sweden. I should like to thank Ulf Hannerz, then the center's codirector, for suggesting that I reexamine classic ethnography to determine whether an anthropological theory of technological activity could be recaptured.

REFERENCES CITED

Barlow, K., and D. M. Lipset

1997 Dialogics of Material Culture: Male and Female in Murik Outrigger Canoes. *American Ethnologist* 24:4–36.

Barton, C. M., G. A. Clark, and A. E. Cohen

1994 Art as Information: Explaining Upper Paleolithic Art in Western Europe. *World Archaeology* 26:185–207.

Bijker, W. E., T. P. Hughes, and T. J. Pinch (editors)

1987 *The Social Construction of Technological Systems: New Directions in the Sociology and History of Technology.* MIT Press, Cambridge, Mass.

Childs, S. T.

1991 Iron as Utility or Expression. In *Metals in Society: Theory beyond Analysis,* edited by R. M. Ehrenreich, pp. 33–46. MASCA Research Papers in Science and Archaeology No. 8(2). University of Pennsylvania, Philadelphia.

Damon, F. H.

1989 *From Muyuw to the Trobriands: Transformations along the Northern Side of the Kula Ring.* Northern Illinois University Press, De Kalb.

David, N. C., J. Sternery, and K. Gavua

1988 Why Pots Are Decorated. *Current Anthropology* 29:365–379.

DeMarrais, E., L. J. Castillo, and T. Earle

1996 Ideology, Materialization, and Power Strategies. *Current Anthropology* 37(1):15–31.

Dobres, M-A.

1995a Gender and Prehistoric Technology: On the Social Agency of Technical Strategies. *World Archaeology* 27(1):25–49.

1995b *Gender in the Making: Late Magdalenian Social Relations of Production in the French Midi-Pyrénées.* Ph.D. dissertation, Department of Anthropology, University of California at Berkeley. University Microfilms, Ann Arbor.

1996 Variabilité des activités Magdaléniennes en Ariège et en Haute-Garonne, d'après les chaînes opératoires dans l'outillage osseux. *Bulletin de la Société Préhistorique Ariège-Pyrénées* 51:149–194.

Dobres, M-A., and C. R. Hoffman

1994 Social Agency and the Dynamics of Prehistoric Technology. *Journal of Archaeological Method and Theory* 1(3):211–258.

Gosselain, O. P.

1992 Technology and Style: Potters and Pottery among the Bafia of Cameroon. *Man* 27:559–586.

Hannerz, U.

1993 *Cultural Complexity: Studies in the Sociological Organization of Meaning.* Columbia University Press, New York.

Hosler, D.

1996 Technical Choices, Social Categories, and Meaning among the Andean Potters of Las Animas. *Material Culture* 1:63–91.

Lechtman, H.

1977 Style in Technology: Some Early Thoughts. In *Material Culture: Styles, Organization, and Dynamics of Technology*, edited by H. Lechtman and R. S. Merrill, pp. 3–20. West Publishing Co., St. Paul, Minn.

Lemonnier, P.

1983 L'Etude des systèmes techniques, une urgence en technologie culturelle. *Techniques et Culture* 1:11–34.

1986 The Study of Material Culture Today: Toward an Anthropology of Technical Systems. *Journal of Anthropological Archaeology* 5:147–186.

1989 Bark Capes, Arrowheads, and the Concorde: On Social Representations of Technology. In *The Meaning of Things: Material Culture and Symbolic Expression*, edited by I. Hodder, pp. 155–171. Unwin Hyman, London.

Leroi-Gourhan, A.

1943 *Evolution et techniques: L'homme et la matière.* Albin Michel, Paris.

1945 *Evolution et techniques: Milieu et techniques.* Albin Michel, Paris.

Lewis, G. R.

1908 *The Stanneries: A Study of the Medieval Tin Miners of Cornwall and Devon.* Bradford Barton, Truro (1965 reprint).

Malinowski, B.

1922 *Argonauts of the Western Pacific.* Routledge and Kegan Paul, London.

1935 *Coral Gardens and Their Magic: A Study of the Methods of Tilling the Soil and of Agricultural Rites in the Trobriand Islands.* Vols. 1 and 2. American Book, New York.

1954 *Magic, Science, and Religion, and Other Essays.* Doubleday Anchor, New York.

Mauss, M.

1935 Les techniques du corps. In *Sociologie et psychologie, Parts II–VI.* Reprinted in *Sociologie et anthropologie*, pp. 365–386. Presses Universitaires de France, Paris, 1950; also in *Sociology and Psychology: Essays of Marcel Mauss*, translated by B. Brewster, pp. 97–123. Routledge and Kegan Paul, London, 1979.

Miller, D.

1987 *Material Culture and Mass Consumption.* Basil Blackwell, New York.

Munn, N. D.

1974 Spatiotemporal Transformations of Gawa Canoes. *Journal de la Société des Océanistes* 33:39–52.

1986 *The Fame of Gawa: A Symbolic Study of Value Transformation in a Massim (Papua New Guinea) Society.* Cambridge University Press, Cambridge.

Pfaffenberger, B.

1992 Social Anthropology of Technology. *Annual Review of Anthropology* 21:491–516.

Prown, J. D.

1982 Mind in Matter: An Introduction to Material Culture Theory and Method. *Winterthur Portfolio* 17:1–19.

Rule, J.

1992 A Configuration of Quietism? Attitudes towards Trade Unionism and Chartism among the Cornish Miners. *Tidschrift voor Sociole Geschiedenis* 18:248–262.

Forthcoming A Risky Business: Death, Injury, and Religion in Cornish Mining, c. 1780– 1870. In *Social Approaches to an Industrial Past: The Archaeology and Anthropology of Metalliferous Mining,* edited by A. B. Knapp, V. C. Pigott, and E. W. Herbert. Routledge, London.

Schiffer, M. B., and J. M. Skibo.

1997 The Explanation of Artifact Variability. *American Antiquity* 62:27– 51.

Sillitoe, P.

1988 *Made in Niugini: Technology in the Highlands of Papua New Guinea.* British Museum, London.

Tambiah, S. J.

1990 *Magic, Science, Religion, and the Scope of Rationality.* Cambridge University Press, Cambridge.

van der Leeuw, S. E.

1993 Giving the Potter a Choice: Conceptual Aspects of Pottery Techniques. In *Technical Choices: Transformation in Material Cultures since the Neolithic,* edited by P. Lemonnier, pp. 238–288. Routledge, London.

III WORLD VIEWS AND TECHNOLOGY

8. Dogs, Snares, and Cartridge Belts: The Poetics of a Northern Athapaskan Narrative Technology

ROBIN RIDINGTON

THE WORD "TECHNOLOGY" HAS COME to be understood as a synonym for artifact rather than for the artifice used in the production of material objects. In recent popular usage, technology refers to objects manufactured through complex industrial systems of production. "High tech" evokes artifacts such as computer chips and lasers. "Low tech" evokes snares or bows and arrows. But the word's root meaning, *technē*, refers to something closer to technique or performance. This chapter describes the performance, poetics, and narrative technology of one hunting people, the Dunne-za Athapaskans (Beaver Indians) of the Canadian subarctic (see figure 8.1). It suggests that their technology is a form of artifice; it is not the artifacts that artifice produces (Ridington 1983; see also Dobres chapter 6 and Pfaffenberger chapter 7 in this volume). Lechtman (1977:12–13) makes a similar point about technology:

> Artifacts are the products of appropriate cultural performance, and technological activities constitute one mode of such performance. . . . It is the synthesizing action of the style, the rendering of the performance, that constitutes the cultural message. Technologies are performances; they are communicative systems, and their styles are symbols through which communication occurs.

Lemonnier, following Mauss, similarly defines technology as action, but he cautions that it "needs to involve at least some physical intervention which leads to a real transformation of matter" (Lemonnier 1992:5; discussion in Dobres chap-

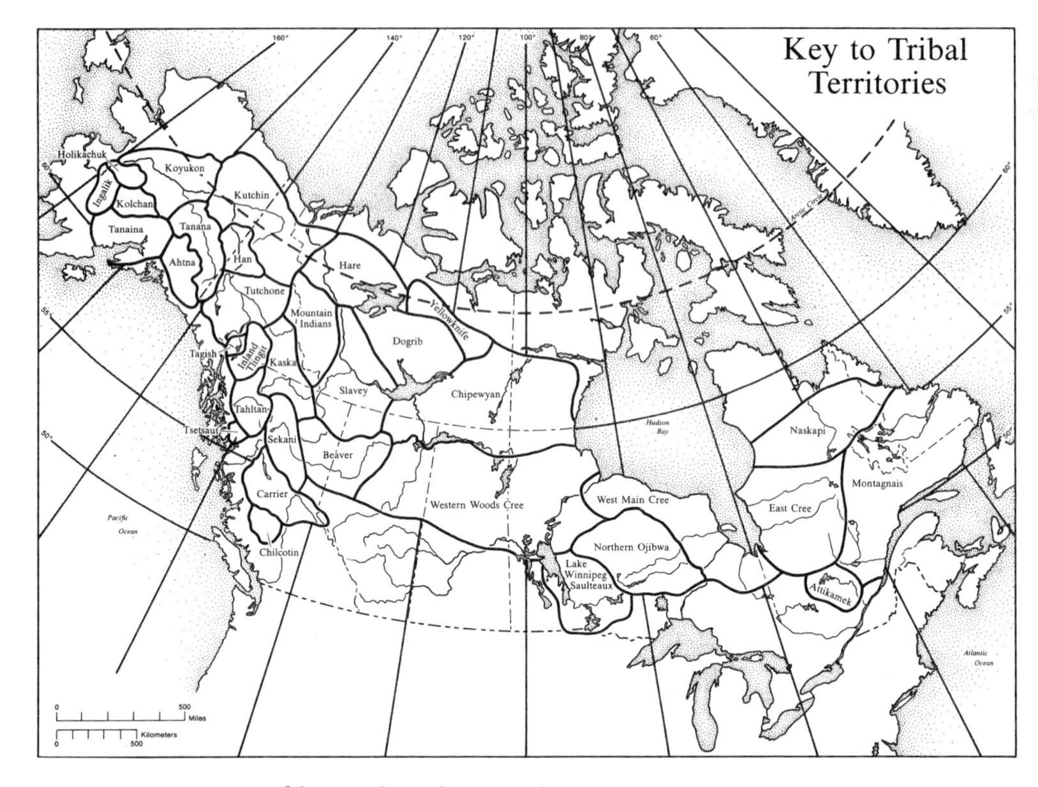

Figure 8.1. Map of the Canadian subarctic (Helm 1981, p. ix; reprinted with permission).

ter 6 in this volume). The material connection he points to is not just the artifacts produced by action; it is the entire realm of interaction between humans and the physical world (see Ingold's Foreword in this volume). Technology is, he says, the application of "specific knowledge" to matter, energy, and objects. By specific knowledge, he means "the end result of all the perceived possibilities and the choices, made on an individual or societal level, which have shaped that technological action" (Lemonnier 1992:6). Thus technology is a kind of economizing behavior that can be found within any sort of economy and is amenable to the interpretations of a formalist economic anthropology.

From the performance/action perspectives of Lechtman and Lemonnier, hunter-gatherer technology is not necessarily a lower form than that of an industrial society. Both systems use artifice as a means of interacting with the material environment. Both invoke knowledge and performance. Both involve strategic decision making. The obvious difference, of course, lies in the sheer size and complexity of the artifacts produced. Perhaps that difference explains why in contemporary popular usage, technology has come to mean artifact rather than the artifice behind it. As "artifactual chauvinists" conditioned to a world that is saturated with material products, we are inclined to see technology as artifact rather than as an underlying interaction with the physical world. We see products rather than process. We may, indeed, be at risk as a species because of the way our dazzling artifacts blind us to the material limits of a finite global ecology.

Ingold (1993:438) suggests that "technology" has come to mean artifact rather than artifice in part because productive forces have been alienated from social life:

> In hunting and gathering societies, the forces of production are deeply embedded in the matrix of social relations . . . the "correspondence" between technical forces and social relations is not external but *internal,* or in other words, the technical is one *aspect* of the social. The modern semantic shift from technique to technology, associated with the ascendance of the machine, is itself symptomatic of the disembedding of the forces of production from their social matrix.

In comparing hunter-gatherer technology with that of industrial societies, Ingold points to a transformation of "the entire system of relations between worker, tool, and raw material." Such a transformation replaces the "subject-centered knowledge and skills" of hunter-gatherers with "objective principles of mechanical functioning." It reflects an evolutionary "objectification" rather than a "complexification" of productive forces (Ingold 1993:439). As Marx has argued, and Charlie Chaplin so eloquently demonstrated in his classic film *Modern Times,* workers in complex industrial systems are objectified and commodified. Workers in post-industrial systems may suffer the even worse indignity of being "decommodified" and removed from the technological loop entirely.

Although all technology may be viewed as being knowledge-based, the techniques with which people in hunting economies relate to one another and to their environment are particularly dependent on knowledge held by individuals and communicated through oral tradition. Hunting people, both men and women, maintain intimate physical and interpersonal relations with the animal people of their environment. Humans and animals are principal characters in stories that define their relations to one another. Their material world is also a storied world. Communication within a matrix of social relations that includes relations with animal people is central to the forces of production in a hunting economy.

It is important to remember that although hunter-gatherer "forces of production are deeply embedded in the matrix of social relations," these relations are themselves embedded in the material and ecological conditions of a natural environment, as Lemonnier suggests. A significant ecological fact of the hunter's environment is that the lives of animals are autonomous and independent from those of humans. The "subject-centered knowledge and skills" of hunter-gatherer epistemology includes an understanding that animals are also willful and subjective beings. Hunting technology is based on the premise that in order to be successful, the hunter must negotiate a relationship with his game. Relations with these beings are essentially interpersonal relations. Hunters may persuade and even coerce animals, but they do not own or control them as do people in economies with domesticated animals. Animals behave as they do for their own reasons, not for the benefit of humans.

Although hunters certainly do alter habitat and influence animal population levels, they do not control how, when, and where animals physically reproduce themselves. They do not control how and where animals make a living. Hunters understand that animals are sentient creatures with distinctive purposes and personalities. They view animals as fundamentally undomesticated persons with whom they must negotiate mutually beneficial relationships. Animals are like humans in being persons, but they are also fundamentally different in being nonhuman persons. Hunters believe that animals choose when and how to encounter a hunter, but they also believe that humans have the capacity to influence that choice. Thus, human/animal interpersonal relationships are sometimes as complex and ambivalent as those that exist between humans. Brightman (1993:188) notes that the Algonquian Rock Cree appear to hold different models of the relations between human and animal persons:

> Some Crees say that hunting is possible *only* with the permission of the animals or game rulers . . . some Crees say that it is *itatisiwak,* or "natural," for animals to avoid hunters, and the gestures of respect are intended to overcome this "natural" disposition and dispose them favorably to the event of their deaths.

Rock Crees sometimes go even farther, when they "talk about hunting in terms that represent animals as opponents or reluctant victims and killings as domination rather than reciprocity" (Brightman 1993:190). Brightman concludes that Crees hold conflicting "benefactive" and "adversarial" ideologies of hunting, but it may also be the case that the two models relate to different phases of the hunting enterprise. Brightman's own data provide a possible resolution. He reports that compulsive measures usually precede a hunt, while expressions of gratitude follow its successful completion. In my own study of the Dunne-za medicine fight (Ridington 1968), I point out that apparently contradictory theories formulated to explain misfortune are, in fact, situationally determined. A person to whom misfortune has happened blames others; a person who feels he is under suspicion blames the victim. Brightman's study indicates that, in a similar manner, relations between humans and animal persons are complex and situationally defined in ways that parallel relations between humans.

NARRATIVE TECHNOLOGY IN A HUNTING ECONOMY

Wendell Oswalt dedicates his book on the evolution of hunting technology "To the maker of man—THE STICK" (Oswalt 1973:v). I suggest that a more appropriate maker of what it is to be human should be the story. Oswalt classifies and ranks material "technocultures" according to the number of material components or "technounits" that comprise a particular artifact. Among ethnographically known hunting cultures, he ranks Inuit technology as the most complex and that of aboriginal Tasmanians as the simplest. By reversing Oswalt's emphasis from the material artifact in isolation, to the social artifice within which tool use is embedded, it is possible to understand a technology in terms of its overall *strategic* complexity rather by the complexity of material artifacts within a particular toolkit.

Strategic complexity in a hunting society should perhaps be thought of as a measure of how men and women apply their social skills to the opportunities and limitations of the environment, rather than as a measure of their material toolkit in isolation from how they deploy it (see also Dobres chapter 6 in this volume). Such a notion has, of course, considerable import for archaeology. If complexity resides in their stories, not their sticks (and stones), then archaeologists must interpret physical objects in relation to an immaterial narrative technology that is not immediately available to their observation. They must practice what Davis (1989:204) calls an "archaeology of thought," bearing in mind that although "not all archaeological remains are remains of mental life, . . . all mental life has remains." Among contemporary hunter-gatherers, some of those remains are coded in the form of oral tradition and performance.

Unlike people trying to cope with a global industrial economy, hunters enjoy a close feedback of information about the efficacy of human action in relation to environmental constraints and opportunities. Their actions are unlikely to cause long-term feedback hazards such as global warming, overadaptation to nonrenewable resources, and the depletion of stratospheric ozone, although they may generate short or medium-term feedback hazards such as game depletion. Hunters make decisions about hunting strategy, band movement, and band size in relation to the immediate feedback of information from the environment and with reference to cultural information communicated through narrative.

Because hunting people cultivate knowledge about seasonal and cyclical variations in their environment, they may actually be less tied to environmental constraints than are food producers. They are strongly motivated to pay serious attention to what is going on around them, and what is going on around them provides information relevant to immediate strategic decisions. Their narrative technology reinforces what Winterhalder and Smith (1981) identify as an "optimum foraging strategy." Optimum foraging requires strategies based on information about periodic fluctuations in resource availability. "Highly efficient short-term behaviors may not be effective over the longer period in which organismal or habitat variability occurs" (Winterhalder and Smith 1981:18). "Efficient and effective foraging behavior [is essential for] the maintenance of viable populations" (Winterhalder and Smith 1981:20). Narrative technology can thus be seen as an enabling instrument of an optimization strategy.

Strategies of resource scheduling and seasonality require that people share and internalize common understandings about long and short-term environmental conditions. Stories are integral to the operation of such a technology. The complexity of hunter-gatherer technology lies in the discourse of its "subject-centered knowledge." The subjects who participate in this discourse include both human and nonhuman persons. Among northern hunting peoples, the technology of discourse with nonhuman persons is in the possession of individuals, not institutions. Individual knowledge about how to communicate with a world of sentient human and nonhuman persons is integral to the adaptation of aboriginal North American hunters, as it must have been to that of the Upper Palaeolithic people who are ancestral to all modern humans. Whenever and wherever the conditions of a hunting economy have prevailed, humans must certainly have viewed the undomesticated animals upon whom they depended as powerful and sentient beings like themselves. Although this phenomenology is not reflected directly in the archaeological record, it is relevant to an understanding of it.

The technology of people in a northern hunting economy requires a particularly sophisticated interaction with a complex natural ecosystem. It requires negotiating relations with sentient animal persons, and it requires communicating information about these relations with fellow humans. Among hunting people of the Canadian subarctic, a person who has established a special relationship to the storied world that human and animal persons share is said to "little bit know something" (Ridington 1990). Such knowledge is regarded as a token of the power a person has gained from his or her childhood vision quest experience.

An empowering system of knowledge gives life to northern hunting people, as it does to all culturally modern humans. Among northern hunters, narrative technology is a way of communicating and demonstrating knowledge. Knowledge explains their shamanic cosmologies and practices, as well as the reciprocities they practice with one another and with the environment. Hunting technology, in order to be performed, must also be communicated between individuals and between generations. Storytelling (discourse and narrative) is the medium through which people communicate knowledge from one person to another and from generation to generation. Storytelling is probably the master trope of our species. Landau (1984, 1991) and Terrell (1990) have written that even the presentation of scientific information takes the form of narrative. As Gould (1994:26) recently pointed out:

> We are storytelling creatures and should have been named Homo narrator (or perhaps Homo mendax to acknowledge the misleading side of tale telling) rather than the often inappropriate Homo sapiens. The narrative mode comes naturally to us as a style for organizing our thoughts and ideas.

The world view, cosmology, and narrative performance of northern Athapaskans are integral to the enabling technology of their cultural ecology. Their social relations with one another and with the nonhuman persons of their environment are also technological relations. Dobres and Hoffman (1994:216) summarize technological relations as "the complex webs interconnecting the material with the social, political, economic, and symbolic experiences of human existence." They identify in the academic literature two "primary approaches to the social dimensions of technology." The first "views technology as an expression of world views"; the second "highlights dynamic social processes of technological activity" (Dobres and Hoffman 1994:216). In their view, an integrated approach has not yet been developed. Northern Athapaskan narrative technology is both an expression of world view and an example of dynamic social processes. By looking at the artifice of northern Athapaskan storied lives, I hope to bring together these two perspectives.

DOGS, SNARES, AND CARTRIDGE BELTS: A DUNNE-ZA STORY ABOUT TECHNOLOGY

The Dunne-za (Beaver Indians) are Athapaskan hunters of the Peace River area. During the course of my first fieldwork with them in the 1960s, I was fortunate to document a wealth of stories from men and women who had spent most of their lives in the bush. One of these was a prophet, or Dreamer, named Charlie Yahey. I recorded his stories in the Beaver language and then worked with younger bilingual translators to produce written English texts.

It is notable that when northern Athapaskans encountered items of European material culture for which they had no cultural context, they responded by dreaming them into their oral history. The oral traditions of many subarctic Athapaskans tell of prophets who foretold the coming of white people and their artifacts (Cruikshank 1994; Moore and Wheelock 1990; Ridington 1987). This tradition of prophecy can be understood as the way these people contextualized a foreign material technology within the technology of their narrative tradition. In Cruikshank's (1994:163) words, prophecy narratives "may be viewed as successful engagement with changing ideas."

Charlie Yahey was a prophet whose stories demonstrate that successful engagement. He narrated a series of stories about Dunne-za philosophy and epistemology. At my request, he told a version of the "earth diver" story about how the germ of substance was brought up from beneath the water by a diving muskrat.

> I guess you heard that story before but you want to hear it twice to see
> which one is better.
> He made this world and at first there were no animals.
> There was just water and no land.
> Then he started to make the land.
> He finished all the land.
> Finally this world started to move
> started to grow and kept growing.
> That is what the old people said.
> Just the water and no land.
> There were no animals.
> Where are they going to stay with no land?
> Only God stayed someplace where he made it for himself
> Maybe boat or just water.
> No land.
> There was just water
> and God made a big cross that he floated up on the water.
> He floated the cross on the water.
> He floated that cross on the water
> and then he called all the animals that stay in the water.
> He sent them down to get the dirt but they just came out.
> They couldn't get it. Too far down.
> The last one was rats [muskrat].
> He sent him down to get the dirt
> and he stayed down for how long.
> Finally he just brought up a little dirt.
> He put that little piece of dirt on the cross and told it
> "You are going to grow."
> From there it started to grow and kept on growing
> every year like that.
> Finally it was getting bigger and pretty soon it was big.
> That is what the old people say.
> Then he made a dog for himself—his own dog.
> He said to that dog
> "You go around to see how big that world is.
> Then you come back here."
> He started off—that dog—to go around the world
> circling around the edge of the water

and when he came back he had a person's bone in his mouth.
Some of the animals on the land people couldn't eat
and he just sent them down under the earth.
Those animals that people cannot live from
he sent them down.
The second time he made a wolf.
He made him out of dog.
He made his dog first and then he threw that dog away
after it came back with a person's bone.
He didn't like that so he got rid of him and got the wolf.
You know some animals he made them do wrong.
Even that woman he [*sic*] do wrong.
He stole the berries.
So animals he made that do wrong he got rid of them.
The devil took them.
Even us—we do wrong he gets rid of us.
He made us but he gets rid of us. We go the other way.
That is what starts to happen.
That wolf started to go traveling.
Finally he was gone and he never came back.
He never showed up again.
God said,
"I love my dog but I don't know where it is. He is lost."
He really knew it—he knew his wolf—but he just pretended.
He wanted the wolf to live with his teeth
to travel around and kill moose.
He wrote that.
He made the wolf's teeth out of steel
and even today he can grab anything just like with a knife.
His teeth cut right through.
He made this world really big and that wolf got lost.
He wanted that wolf to get lost.
That is why the wolf never came back.
The wolf is going to be on this world too.
He made everything really perfect.

Charlie Yahey portrayed Muskrat in this story as a person who dives down to recover substance at the center of an idea. The story is about more than how the world began. It also explains the dimensionality in which human experience takes place. The center of the cross represents the essential image of the place where two trails come together. Muskrat's dive is a "magical flight," the first part of a shamanic journey to create the world. Muskrat's dive may be seen as an archetypal form of northern hunting technology. By retrieving the first germ of substance from the first idea of intersecting trails, Muskrat brings the world to life. By dreaming the place where his trail crosses that of an animal, the human hunter maintains it.

Once the world had begun to grow, the creator in Charlie Yahey's story wished to measure its ecological self-sufficiency. He sent out a wolf to measure its size, but the wolf came back with a human arm in its mouth. The creator took this to be a sign that the world was not yet a self-sustaining system. The next wolf he made never came back, because it could "live with his teeth to travel around and kill moose."

Charlie Yahey's elaborated version of the widely distributed "earth diver" story sets out some essential features of Dunne-za hunting technology. In the story, the creator imagines a sense of place, using a cross that he floats on the water. Although this image certainly has some resonance with the Christian cross in Dunne-za eyes, it more fundamentally illustrates the essential fact that, to live by hunting, the hunter must visualize a place where his tracks and those of an animal come together. I learned that, in Dunne-za theory, the real hunt takes place in a dream in which the hunter experiences that point of contact. In the dream, the animal gives itself to the hunter. The relationship between hunter and game is interpersonal. Indeed, it may be seen as an essential social relation of production that is, to repeat Ingold (1993:438), "deeply embedded in the matrix of social relations." In this case, social relations include not only relations with other humans, but also relations with a whole range of other nonhuman persons. Again recalling Ingold, "The 'correspondence' between technical forces and social relations is not external but *internal,* or in other words, the technical is one *aspect* of the social" (Ingold 1993:438). Animals and natural forces are, to the Dunne-za and other northern hunting people, sentient subjects, not impersonal objects. The subject-centered knowledge and skills that constitute a northern hunting technology assume a discourse that all such sentient persons share.

Northern hunting cultures are typically flexible, adaptable, and ready to take advantage of variations in the resource potential of their environment. In historic times, they have readily integrated new social forms and items of material culture into their ways of doing things. Part of their social technology has been to incorporate new material goods into the storied world. Charlie Yahey's version of the creation story freely incorporates imported items of material culture into a typically shamanic earth-diver tale. The wolf lives successfully because the creator "made the wolf's teeth out of steel."

Following his account of the earth-diver episode, Charlie Yahey continued with a story about how the Dunne-za view their own adaptive strategy compared with that of the newcomers from Europe. The story relates that, long ago, the creator gave the people a choice as to how they would make a living. One option was to write a design for whatever they wanted on paper and have it come true without further effort. The other, was to make a living using the tools of "dogs, snares, and cartridge belts." The story illustrates both essential features of Dunneza technology and also its adaptability. Here is the story as Margaret Davis translated it:

> God made everything on this world
> by drawing out the design for it
> on a piece of paper.
> He made dogs, snares, and cartridge belts.
> Then, he took these and the paper for drawing designs
> to the people of long ago.
> He put these things before the old men.
> He said,
> "Anything you want from this land
> when I have finished making it
> I will write down on this piece of paper.
> You can choose which gifts you want;
> the paper to make anything you want,
> or the dogs, snares and cartridge belts."
> But the Indians said to the paper,
> "We won't get anything from this piece of paper,"
> and they took the other gifts instead.
> Dogs barking.
> People can live from the dogs.
> When people go to hunt they take the dogs with them
> and the dogs show them where to hunt.
> "From this paper we will get nothing," they said.
> So they took the snares and cartridge belts
> and they knew about them.
> The white people took the piece of paper.
> They can make everything; wagons, stores.
> He wrote it down on that piece of paper for them.
> Even these airplanes he made for them.
> This world is not big enough for them.
> . . .
> He made us Indians to live in the bush,
> to do hard jobs and to make our living.

We just do our own lives,
but the white men started growing crops.
They made plants that would grow and started to copy them.

The white men, Charlie Yahey says, can make everything from a piece of paper, but "this world is not big enough for them." When the wolf came back with a human bone in its mouth, the creator took it as a sign that the world had not yet achieved sustainability. Now, the white people, with their paper-based technology that produces "wagons, stores, and airplanes"—Ingold's "objective principles of mechanical functioning"—have taken the world back to a condition of ecological imbalance. It is "not big enough for them."

Indians, according to Charlie Yahey, chose dogs, snares, and cartridge belts. When I heard Margaret's translation of this story, I was initially surprised and disappointed to hear cartridge belts listed as part of an original Dunne-za technology. Surely, the people of long ago used bows and arrows, not breech-loading rifles. Other stories told clearly of a time, not so long ago, when the Dunne-za first learned about muzzle-loading muskets. Then I asked her what word in Beaver Charlie had used. She replied that it was *atu-ze,* which, she confidently told me, was how you say "cartridge belts" in Beaver. Both she and I knew, of course, that *atu* means "arrow" and *ze* means "real, proper to, or belonging to." Thus, *atu-ze* could be translated literally as something like "belonging to real arrows" and may have once meant either arrow holder or bow. But Margaret insisted that *atu-ze* is "our word for cartridge belts." Suddenly it dawned on me that rather than being an example of cultural contamination and anachronism, this story demonstrates a continued cultural vitality and adaptability. It is about the essence of Dunne-za adaptive strategy and how it differs from that of the white men.

Indians make their living from their knowledge of the environment. They make it through negotiating social relations with sentient nonhuman persons. The particular instruments of this technology are not essential to its successful operation. Once, people used bows and arrows. Now they use rifles and cartridges. The essence of their technology is situated in the mutually understood social relations of production they negotiate with human and nonhuman persons, rather than through the possession of any particular artifact. "Real Indians" are not constrained by the artifactual inventory of their ancestors. Real Indians, Dunne-za, use whatever instrumental extensions of their intelligence are available to them.

In the time of the prophets who first dealt with the white men, communal hunting with snares was a profitable technique. In Charlie Yahey's time, Indians hunted with breech-loading rifles. I often saw the old man setting out to hunt with his classic lever-action Winchester Model 94. More often than not, he came back with fresh meat. Dogs, snares, and cartridge belts symbolize the instruments through which Dunne-za hunters of his generation made contact with the animal persons of their environment. Dunne-za technology is about the application of knowledge to a sentient and interpersonal environment. In Charlie Yahey's words:

> He made us Indians to live in the bush,
> to do hard jobs and to make our living.
> We just do our own lives.

NARRATIVE TECHNOLOGY AND ADAPTIVE STRATEGY

The narrative tradition of hunting and gathering cultures allows them to be remarkably flexible, adaptable, and ready to take advantage of variations in the resource potential of their environment. Knowledge necessary for informed decision making is widely distributed among adult members of small-scale hunting and gathering communities. The egalitarianism found in these communities functions successfully because individuals are expected to be in possession of essential information about their natural and cultural environment (for further discussion, see Dobres chapter 6 in this volume). Discourse within such an oral culture is highly contextualized and based on complex mutually understood (and unstated) knowledge (Brody 1981; Ridington 1988).

Information in an oral tradition is stored in a way that is analogous to the distribution of visual information in a holographic image. Each person retains an image or model of the entire system of which he or she is a part. Each person is responsible for acting autonomously and with intelligence in relation to that knowledge of the whole. Each person knows how to place his or her experience within its meaningful pattern. People experience stories as small wholes, not as small parts of the whole. They are not meaningless components of a coded message analogous to phonemes but, rather, are small examples of a meaningful totality. Their smallest components remain semantic and ideographic. Narrative performance plays creatively upon that mutually understood totality, as does the performance of technology (see similar arguments in Hoffman chapter 5 and Pfaffenberger chapter 7 in this volume).

In the technology of storied experience, each performer's speech and action evoke and are meaningful in relation to everything that is known but, for the moment, unstated. Each story contains every other story. Each person's life is an example of the mythic stories that people know to exist in a time out of time. As in post-structuralist semiotic theory, in which a sign is meaningful in relation to what it is not (Derrida 1976), experience within a closely contexted oral tradition is meaningful in relation to an unstated but mutually understood totality. Storied speech is an example of that totality, not simply a part of it. Like Muskrat's speck of dirt that is still the world, its smallest components are still stories. They make sense as metonyms, parts that are also wholes.

Communication based on assumed mutual understandings has been described as a "restricted" as opposed to an "elaborated" code of discourse (Bernstein 1966). Although Bernstein developed his distinction in reference to class differences within a modern urban culture, his terms may be adapted to describe an important quality of communication within an oral hunting and gathering culture. Elaborated discourse refers to communications in which information is introduced with a summary of the context to which it refers. The speaker "takes nothing for granted from the audience and sets out the context and the dramatis personae in a journalistic mode" (Cruikshank 1985:11). It is adapted to situations in which people who do not share knowledge must establish some common ground for communication. Restricted discourse refers to communications in which the context is taken for granted.

When applied to the oral communication medium of highly individualistic hunting and gathering communities, Bernstein's "restricted" code might better be called a "reflexive" code. His distinction would then refer to differences between systems in which the context is assumed, as opposed to those in which it must be specified. Hall (1983) has described a similar phenomenon in the distinction he makes between "high and low context messages." According to Hall (1983:56–57), "the more information that is shared . . . the higher the context." If a statement's meaning is dependent on the context in which it is made, as Hall argues, people who share common knowledge and experience may be expected to communicate through highly contextualized language.

Hunter-gatherer adaptation depends in particular upon shared knowledge, mutual understanding, and shared codes. A social setting in which the context of communication is assumed provides ample opportunity for metaphors that refer to mutually shared knowledge and experience. Metaphors based on mutual understandings allow for considerable economy and subtlety of communication. They condense experience gained in contexts that are understood mutually and apply this shared knowledge to an individualized set of circumstances. Biesele (1984) has called this process the "multiplier effect" of metaphor. Stories about mythic time make sense because they are condensations of the experience of everyday reality. Their images and meanings are constant and indwelling presences. They inform the conduct of everyday life (see also Pfaffenberger chapter 7 in this volume).

Aboriginal people of the North American subarctic have evolved adaptive strategies that place great emphasis on the authority of individual intelligence within the social responsibility required of a system in which animals and humans alike are interdependent members of a single community. They recognize that success in hunting and other activities depends more on the possession of knowledge and reciprocities with other persons than on the possession of particular material goods. They rely on narrative knowledge in the possession of individuals, rather than on knowledge that is mediated through supra-individual institutions. Unlike more sedentary people who can accumulate wealth in the form of material possessions, subarctic people recognize knowledge as a form of wealth. Physical objects may be lost, but knowledge stays with a person throughout his or her life. Knowledge can be communicated and shared through narrative. Drawing upon narrated knowledge, a person can use environmental resources to make material objects as they are needed at a particular site.

Northern hunters find it far more attractive to carry plans and information in their minds, rather than to be burdened with carrying material artifacts (Ridington 1982). Physical objects entail a high cost when they compete with other physical necessities, such as children, clothing, and trail food, for the very limited carrying capacity of the human body. Various strategies were available to minimize this limitation. One was to cache duplicate kits at strategic locations to which a band returned periodically; another was to limit the material complexity of artifacts to those that could be made quickly of locally available materials. Thus an inventory of simple, easily replaceable, tangible artifacts may actually reflect a complex adaptive strategy, based on the possession of specialized knowledge. Although a person's inventory of material objects may be small, his or her narrative inventory is typically extensive.

From the hunter's point of view, the most valuable technology would be one that required no material encumbrance whatsoever. Material objects have value only as the final material connection in the deployment of a strategy held in mind. The sophistication of a hunting technology may be measured by its cost efficiency in terms of a ratio of its physical weight and bulk to its productivity. From this perspective, a technology that is carried in the mind and coded in narrative tradition, rather than carried in the hand and coded in the form of an artifact, is highly cost-efficient. The essence of hunting technology is to retain and be able to act upon information about potential relationships between people and the natural environment.

The core of a successful hunting technology can be carried from place to place in the mind. Snare hunting is a prime example of such a knowledge-based technology. Whereas the snare, itself, is a simple one-piece artifact, the artifice with which it is deployed reflects a complex understanding of animal behavior and habitat. The snare, unlike a trap or deadfall, is inert. The hunter loads it with information, not energy. The final killing force is supplied by the animal. The hunter's contribution is as much conceptual as it is physical. The snare is almost, but not quite, a pure sign. For the hunter, it is both a sign and a meaningful instrument. For the animal, it is the releaser of an innate biogenetic program. The animal comes of its own volition to where the hunter has set the snare. The snare itself merely directs a biologically given flight reaction into a positive feedback loop ending in the animal's death.

Because in Dunne-za thought a hunt can be completed only after the hunter has negotiated a relationship with his game in a dream, an animal will come to a snare only after such a negotiation has taken place. The hunter's skill lies in his reading of the landscape in relation to his dream encounter with the animal. He must think like the animal in order to set the snare in an appropriate place. In the past, snare hunting often involved coordinating the efforts of several people to drive an animal toward a snare site. According to Dunne-za oral tradition, Dreamers could visualize the pattern of these hunts in their dreams. Unlike the individual hunter, who dreams ahead only on the trail of his own encounter with an animal, the Dreamer "dreams ahead for everybody" (Ridington 1987). The dream in which he visualizes hunters in relation to an animal may be seen as a form of social technology. It represents a plan of action that members of the task force hold in common. It also symbolizes harmony in interpersonal relations. Charlie Yahey said that animals will come close to humans who are close to one another; who sing and dance together.

CONCLUSION

It seems appropriate to end my own narrative about technology among hunting people with a return to the original character of our species, *Homo sapiens,* as people who are "wise and full of knowledge." In this narrative, as in mythic time, beginning and end are inherent in every moment of experience. We began as sapient hunters and gatherers in possession of a narrative technology. Elements of that past are with us still. Whatever narrative traditions may guide us as individuals, we each carry within us a speck of the world that Muskrat brought up from the primordial ocean of our beginnings as a species. Meaningful culture is still being created through the exercise of individual intelligence. As Gould observed in thinking about our evolutionary history, life within a storied world did not die with the rise of civilization and its "great" religious and scientific traditions. Landau and Terrell remind us that storytelling is as important to modern science as it has been to the lives of hunting people. It remains essential as we look toward the uncertainties of an adaptively unstable global cultural system.

The world of nature is still alive with meaning. Narrative traditions of northern hunting people continue to offer us insights into the possible connections between knowledge, power, and experience. They continue to ground us in the reality of our place within a planetary ecology. We need the intelligence of our hunting ancestors and their contemporary descendants now as we never have before. Our civilization has brought into existence monstrous powers of self-destruction through what Ingold (1993:439) referred to as "the objectification of productive forces." Our cultural intelligence has lost touch with the sapient intelligence we all have as members of the species that is supposed to be "wise and full of knowledge." Although we possess exponentially increasing amounts of information, we do not necessarily use it wisely. Being full of information is not the same thing as being full of knowledge. By objectifying technology, we may have alienated knowledge from experience. Knowing about and knowing how are not the same thing.

Despite these changes, spirits of our hunter-gatherer ancestors live, as we continue to live on the planet they left for us. We continue to be responsible for living the stories that will give life to the generations that follow. As Charlie Yahey said, "The white people took the piece of paper. . . . They can make everything; wagons, stores, airplanes." The only problem, he said, is that "this world is not big enough for them." We can make everything, but we cannot make the world bigger than it is. In order to survive, our artifice must reduce the artifactual burden we lay on the planet's resources. Our "transformation of matter," to use Lemonnier's term, must optimize existing planetary cycles of possibility. "Highly efficient short-term behaviors," as Winterhalder and Smith (1981:18) said of optimal foraging, "may not be effective over the longer period in which organismal or habitat variability occurs." Humans today are far more vulnerable to habitat variability over time than were our hunting and gathering ancestors, in that our habitat is an entire global ecosystem. The story continues.

REFERENCES CITED

Bernstein, B. B.

1966 Elaborated and Restricted Codes: An Outline. In *Explorations in Sociolinguistics*, edited by S. Lieberson, pp. 126–133. Mouton, The Hague.

Biesele, M.

1984 How Hunter-Gatherers' Stories "Make Sense": Semantics and Adaptation. Paper presented at the 1984 Canadian Ethnology Society session, "Information, Imagination, and Adaptation among Hunting and Gathering People." Université de Montreal, Québec.

Brightman, R.

1993 *The Grateful Prey: Rock Cree Human-Animal Relationships*. University of California Press, Berkeley.

Brody, H.

1981 *Maps and Dreams*. Douglas and McIntyre, Vancouver.

Cruikshank, J.

1985 Approaches to the Analysis of Oral Tradition: An Annotated Bibliography. MS on file. Department of Anthropology, University of British Columbia, Vancouver.

1994 Claiming Legitimacy: Prophecy Narratives from Northern Aboriginal Women. *American Indian Quarterly* 18(2):147–168.

Davis, W.

1989 Towards an Archaeology of Thought. In *The Meanings of Things: Material Culture and Symbolic Expression,* edited by I. Hodder, pp. 202–209. Unwin Hyman, London.

Derrida, J.

1976 *Of Grammatology.* Translated by G. C. Spivak. Johns Hopkins University Press, Baltimore.

Dobres, M-A., and C. R. Hoffman

1994 Social Agency and the Dynamics of Prehistoric Technology. *Journal of Archaeological Method and Theory.* 1(3):211–258.

Gould, S. J.

1994 The Neanderthals and Us. *New York Review of Books* 41(17):24–28.

Hall, E. T.

1983 *The Dance of Life: The Other Dimension of Time.* Anchor/Doubleday, Garden City, N.Y.

Helm, J. (editor)

1981 *Subarctic.* Handbook of North American Indians, vol. 6, W. C. Sturtevant, general editor. Smithsonian Institution, Washington, D.C.

Ingold, T.

1993 Technology, Language, and Intelligence: A Reconsideration of Basic Concepts. In *Tools, Language, and Cognition in Human Evolution,* edited by K. R. Gibson and T. Ingold, pp. 449–472. Cambridge University Press, Cambridge.

Landau, M.

1984 Human Evolution as Narrative. *American Scientist* 72:262–268.

1991 *Narratives of Human Evolution.* Yale University Press, New Haven, Conn.

Lechtman, H.

1977 Style in Technology: Some Early Thoughts. In *Material Culture: Styles, Organization, and Dynamics of Technology,* edited by H. Lechtman and R. S. Merrill, pp. 3–20. American Ethnological Society, St. Paul, Minn.

Lemonnier, P.

1992 *Elements for an Anthropology of Technology.* Anthropological Papers No. 88. Museum of Anthropology, University of Michigan, Ann Arbor.

Moore, P., and A. Wheelock

1990 *Wolverine Myths and Legends: Dene Stories from Northern Alberta.* University of Nebraska Press, Lincoln.

Oswalt, W. H.

1973 *Habitat and Technology: The Evolution of Hunting.* Holt, Rinehart and Winston, New York.

Ridington, R.

1968 The Medicine Fight: An Instrument of Political Process among the Beaver Indians. *American Anthropologist* 70(6):1152–1160.

1982 Technology, World View, and Adaptive Strategy in a Northern Hunting Society. *Canadian Review of Sociology and Anthropology* 19(4):469–481.

1983 From Artifice to Artifact: Stages in the Industrialization of a Subarctic Hunting People. *Journal of Canadian Studies* 18(3):55–66.

1987 From Hunt Chief to Prophet: Beaver Indian Dreamers and Christianity. *Arctic Anthropology* 24(1):8–18.

1988 Knowledge, Power and the Individual in Subarctic Hunting Societies. *American Anthropologist* 90:98–110.

1990 *Little Bit Know Something: Stories in a Language of Anthropology.* University of Iowa Press, Iowa City.

Terrell, J. E.

1990 Storytelling and Prehistory. *Archaeological Method and Theory* 2:1–29.

Winterhalder, B., and E. A. Smith

1981 *Hunter-Gatherer Foraging Strategies: Ethnographic and Archaeological Analyses.* University of Chicago Press, Chicago.

9. Exploitation of Tradition: Bone Tool Production and Use at Colony Ross, California

THOMAS A. WAKE

THIS CHAPTER EXAMINES THE sociohistorical significance of the persistence of precontact Native Alaskan bone tool technology at the nineteenth-century Russian-American Company (RAC) fur trading and agricultural outpost of Colony Ross. Recent archaeological investigations at Fort Ross State Historic Park, Sonoma County, California (figure 9.1), have recovered a wide variety of Native Alaskan artifacts and food remains from a living area termed the Native Alaskan Neighborhood (NAN) (Lightfoot, Schiff, and Wake 1997; Lightfoot, Wake, and Schiff 1991, 1993; Mills and Martinez 1997; Wake 1994, 1995, 1997a, 1997b, 1997c). To date, two areas within the NAN have been investigated: the Native Alaskan Village Site (NAVS), and the Fort Ross Beach Site (FRBS) (see figure 9.2).

The analysis of what may be considered relatively banal yet fundamental technological aspects of North Pacific maritime culture will show that bone artifact production certainly has a great deal of anthropological interest above and beyond the material technical production process (Dobres and Hoffman 1994). In fact, the production of bone tools in traditional precontact styles at Ross harkens to Lechtman's (1977) concept of technological style: that is, that "technological performance was supported by a set of underlying values" (Lechtman 1977:10; Dobres and Hoffman 1994:218). The physical production of bone hunting implements at Ross, as with the other case studies in this volume, is a prime example of "how prehistoric technology reaffirmed the very normative values and practices that simultaneously structured technology, in a recursive and dynamic manner" (Dobres and Hoffman 1994:218). The manufacture of traditional precontact bone hunting tools at Ross in the face of profound European technological influences served, at least in part, to reaffirm Native Alaskan cultural identities.

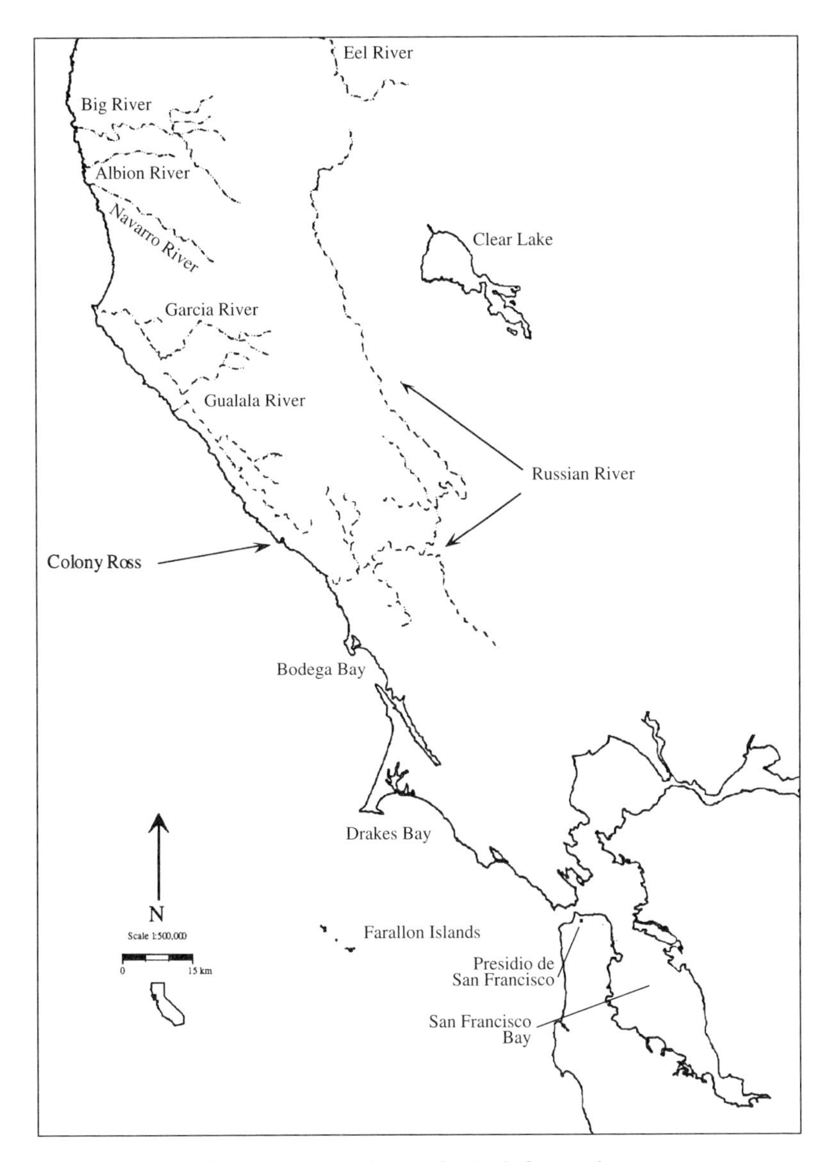

Figure 9.1. Location of Colony Ross (redrawn after Lightfoot et al. 1991).

HISTORICAL BACKGROUND

From its inception, the Russian-American Company exploited the functional efficiency of traditional Native Alaskan culture, technology, and hunting and craft skills for economic gain (Farris 1989; Tikhmenev 1978; Veltre 1990). A review of historical documents combined with archaeological findings indicate that the ma-

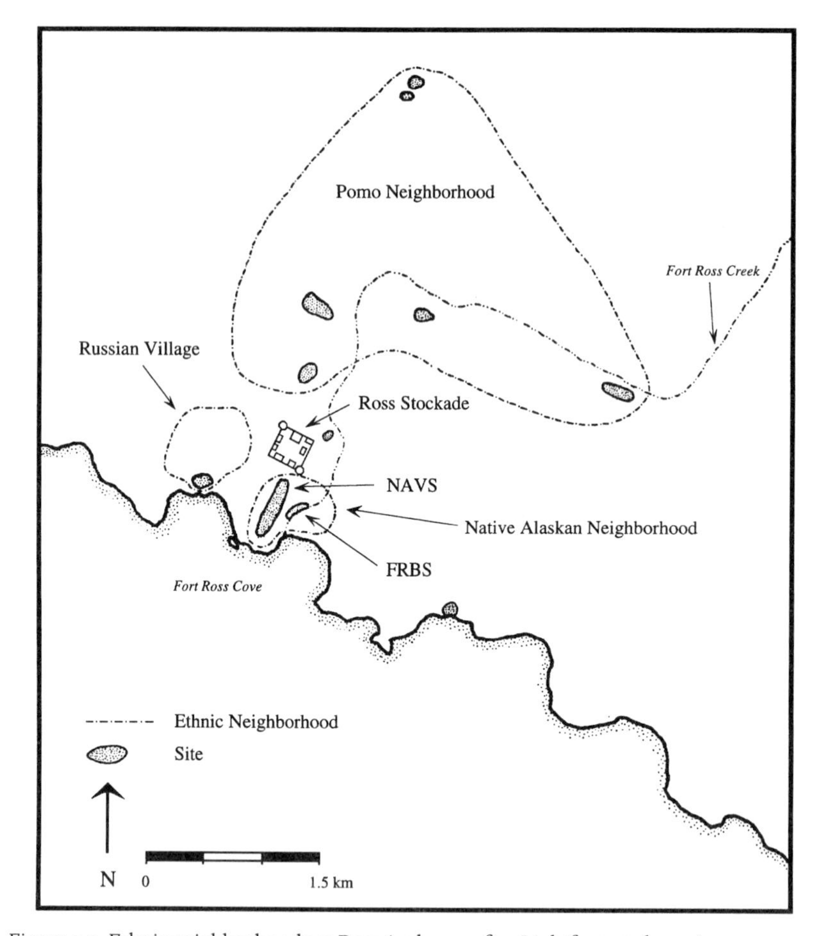

Figure 9.2. Ethnic neighborhoods at Ross (redrawn after Lightfoot et al. 1991).

jority of sea otter hunting at Ross and along the California coast was conducted by Native Alaskan men using time-tested traditional technologies (Khlebnikov 1976, 1990; Lightfoot, Schiff, and Wake 1997; Lightfoot, Wake, and Schiff 1991, 1993; Ogden 1941; Tikhmenev 1978; Wake 1995a, 1997c). As this chapter shows, the RAC had a vested interest in maintaining, at the very least—if not outright encouraging—the continued practice of specific traditional Native Alaskan techniques that provided the company with valuable sea otter pelts. Furthermore, the RAC's conscious exploitation of traditional Alaskan bone tool technology played an important role in maintaining aspects of Aleut and Qikertarmiut (Koniag) cultural identity in the face of European domination. In the late eighteenth and early nineteenth centuries, sea otter pelts were highly desired in China, where the skins were traded for consumer goods such as porcelains and tea, then in great demand in Europe (Chevigny 1965; Dmytryshyn and Crownhart-Vaughan 1988; Dmytryshyn et al. 1989; Okun 1951; Tikhmenev 1978). Sea otter pelts were worth in excess of 250 rubles each during this period, a sum greater than a skilled craftsman's yearly salary of 120 to 200 rubles (Khlebnikov 1990:64).

Two broadly defined ethnic groups at Ross may have benefited from the continuation of traditional Native Alaskan marine mammal hunting practices. Although it is clear that Native Alaskan peoples were exploited and often tragically mistreated by Russian colonists (Black 1977, 1988; Dauenhauer and Dauenhauer 1990; Jacobs 1990; Kari 1990; Pierce 1988), as were the local Kashaya Pomo on at least a few notable occasions (Wrangell 1969), the RAC was not in the same business as the Spanish missions a short journey down the coast from Ross. The Spanish system's missions revolved around a concentrated effort to destroy the traditional identities of native people and forcibly convert them to Catholicism and a European version of a peasant's way of life (Hoover 1985). In contrast, the only known treaty between a European power and Native Californians in the pre-American period was negotiated at Ross in 1817. This treaty was essentially a mutual nonaggression pact that allowed the RAC to establish a settlement on donated land explicitly recognized as originally *belonging* to the Kashaya Pomo (Spencer-Hancock and Pritchard 1981). The result, probably not lost on the Native Californian side, placed the Kashaya Pomo outside the realm of direct influence and potential missionization that their neighbors to the south were forced to endure.

Clearly, the motives of these two European powers were different. The RAC was principally interested in economic gain, and investors in the RAC generated immense profits from their end of the fur trade. However, these profits would have been much harder to come by had the company followed the Spanish example of complete sublimation of traditional culture. The RAC quickly learned that traditional maritime Alaskan hunting and transport techniques were the most efficient, and therefore the most profitable for their interests. These Native Alaskan marine mammal hunters, first encountered by Russian explorers (*Promyshlenniki*) during the early eighteenth century in the Aleutian Islands, and later on Kodiak, were the finest, most efficient sea otter hunters known. The first *Promyshlenniki* and company representatives to make contact with Aleuts and Koniags soon realized, in comparison with these indigenous hunters, how inept they were in the stealthy, demanding hunting practices necessary to take sea otters in great numbers.

Ironically, the RAC's forcible exploitation of the advanced marine mammal hunting technology traditionally practiced by the native Alaskans may have, in an interesting twist of fate, become a means for those co-opted individuals to resist the complete loss or sublimation of their identities as Aleut or Qikertarmiut. It is important to recognize that the technology so fundamental to successful marine mammal hunting, which maritime Native Alaskans had developed over thousands of years, transcended mere subsistence; it was an integral part of their entire cultural identity (e.g., Davydov 1977 [1810]; Fitzhugh and Crowell 1988; Mills 1994; Rousselot et al. 1988; see further elaboration of this argument in Ridington chapter 8 in this volume).

Skin boats (*kayaks*, or *baidarkas*), arguably the most important marine mammal hunting implement after the harpoon, served also as the primary means of immediate transport for precontact Native Alaskans. Somewhat ironically, these boats were also employed by the RAC throughout most of Alaska. The traditional clothing worn while hunting in *baidarkas,* the waterproof seal gut *kamleika,* was also the outergarment of choice worn throughout the often rainy Russian colonies elsewhere in America. Bentwood hunting visors, functionally important headgear while hunting, were decorated with elaborate painted designs, carved ivory plaques and figurines, and beaded seal whiskers. These exotic objects are layered with functionality and symbolism—affording head and eye protection that is wholly functional, while displaying a person's sociocultural identity and mystical requirements for successful hunting.

COLONY ROSS

Colony Ross was founded in 1812 by Ivan Kuskov, an agent of the Russian-American Company, a mercantile monopoly granted by Czar Alexander in 1799. Ross served principally as a base for extended sea otter hunting operations along the west coast of North America, from near the current Oregon border to the tip of Baja California (figure 9.1). The colony's agricultural mission was originally secondary to the immediate profit motives and political aspirations of the RAC; it developed to its greatest extent well after the sea otter was nearly extirpated in California. Ross was short-lived as a mercantile enterprise and was sold to John Sutter in 1841 for a sum of US$30,000 (Du Four 1933; Essig 1933; Farris 1990).

The colony itself consisted of a variety of architectural structures and outlying agricultural fields. The focal point of Ross was the large fortified stockade complex containing administrative buildings, warehouses, a chapel, and Russian living quarters. A wide range of building structures lay outside the walls of the stockade, including agricultural storehouses, a windmill, a smithy, a shipyard, and living quarters for the majority of the populace. The shipyard, located down on the beach in Fort Ross Cove, produced four brigs in excess of 160 tons and a variety of smaller watercraft for other purposes and for the Spanish missions (Allan 1997; Khlebnikov 1976, 1990; Lightfoot, Schiff, and Wake 1997; Lightfoot, Wake, and Schiff 1991, 1993). On-site production of metal tools and fittings for these ships was conducted at the colony's smithies (Khlebnikov 1976, 1990).

The area surrounding the stockade can be divided into three ethnically discrete "neighborhoods": Russian, Native Alaskan, and Native Californian (figure 9.2). As defined here, the Russian neighborhood includes the stockade and the area close to the northwest wall of the fort, where a series of Russian-style cottages were located (Farris 1989; Lightfoot et al. 1991, 1993). The Native Alaskan neighborhood includes the Native Alaskan Village Site, placed directly under the watchful eye of the well-armed southerly seven-sided blockhouse, and the Fort Ross Beach Site, down in Fort Ross Cove. Historical sources state unequivocally that inhabitants of NAVS were given free rein to construct a variety of architectural structures in the marine terrace area to the south and west of the blockhouse. These constructions, placed "haphazardly across the landscape," ranged from traditional Native Alaskan–style semisubterranean *barabaras* (Crowell 1988, 1994), to above-ground "houses made of planks," and circular structures of planks and bark reminiscent of Kashaya Pomo huts (Barrett 1952; La Place 1986 [1839]). The Native Californian Neighborhood includes the village of *Me-Ti-Ni,* to the north and east of the stockade, as well as a series of other small habitation sites still further to the north and east.

PEOPLE AND POWER AT ROSS

Russian America, especially Colony Ross, was populated by a wide variety of people from all over the North Pacific Ocean, including Russians, Siberians, Native Alaskans, Native Californians, mixed-race Creoles, and even some Hawaiians (Fedorova 1973, 1975; Istomen 1992; Khlebnikov 1976, 1990; Lightfoot, Schiff, and Wake 1997; Lightfoot, Wake, and Schiff 1991, 1993; Ogden 1941). Socioeconomic status at the colony was determined by the RAC and was based on race. Russians occupied the most powerful offices, Creoles filled some bureaucratic positions, Native Alaskans provided a labor force primarily devoted to marine mammal hunting (but also manual labor), and Native Californians (primarily Kashaya Pomo) provided unskilled manual labor for various tasks (Khlebnikov 1976, 1990; Lightfoot et al. 1991, 1993). All of these groups were paid either in cash or in kind. Most residents of the colony were paid in scrip redeemable only at the RAC store. The majority of the local Native Californians were paid in kind, with rations of food, clothes, or beads and various trinkets. In actuality, few high-ranking ethnic Russians or Creoles were present at the colony at any given time.

Native Alaskan men from Kodiak Island were the most numerically dominant group of non-Californians living at the colony (Istomen 1992; Khlebnikov 1976). The majority of these men were brought to Ross specifically to hunt sea otters, and secondarily to provide manual labor construction and agriculture. Some of the Native Alaskan men brought to Ross were skilled craftsmen trained by Russians in Alaska. The importance of their craft skills was recognized by the RAC, and trades such as carpentry, blacksmithing, and cooperage earned wages between 150 to 200 rubles a year (Khlebnikov 1990:100). Curiously, there appears to be no mention in the available historical documents of people with bone-carving skills present at Ross.

A few Native Alaskan women were also brought to Ross. However, the majority of women at the colony were Native Californian. Most of these local Kashaya Pomo and Coast Miwok women lived with the Native Alaskan men, residing in interethnic households in the Native Alaskan Neighborhood (Istomen 1992; Jackson 1983; Khlebnikov 1976). Khlebnikov (1976) notes that some of the Native Californian wives of Native Alaskan hunters learned to sew "Alaskan-style" waterproof outergarments, or *kamleikas.* Though Russians controlled most aspects of life at the colony, the demographic population of Ross was dominated by Native Americans, both Californian and Alaskan. Only brief descriptions of the Native Americans and their lifeways are recorded in the available historical record. This illustrates the importance that archaeological research at Ross has for understanding the underreported interethnic backbone of the colony (Davydov 1977 [1810]; Farris 1988; La Place 1986 [1839]; Stross and Heizer 1974; Wrangell 1969).

ARCHAEOLOGY AT ROSS

Early archaeological investigations at Fort Ross focused on the Russian presence and on defining the alignments of the original stockade walls and some of the structures within (Treganza 1954). Throughout the 1960s and 1970s, the focus remained on the Russian presence at Ross (Farris 1981, 1989, 1990; Spencer-Hancock and Pritchard 1982). It was not until the late 1980s that the archaeological view of the colony was refocused to look outside the stockade walls. Even then, the bulk of research was still concerned with the Russian presence. Only recently has attention at Ross shifted to the underreported but more numerous Native Alaskan and Native Californian inhabitants of the colony (Lightfoot, Schiff, and Wake 1997; Lightfoot, Wake, and Schiff 1991, 1993; Mills and Martinez 1997).

Excavation in the Native Alaskan Neighborhood has recovered a great variety of archaeological materials, including shell, animal bones, beads, ceramics, glass, metals, and chipped and ground stone (Lightfoot et al. 1997). Among the more intriguing classes of recovered data are the bone tools, worked bone artifacts, and bone debitage (Wake 1995, 1997a). Most of these relate to the continuous on-site production, maintenance, and repair of hunting and fishing implements. Cut marks present on the bone artifacts suggest that metal tools such as knives, and perhaps hatchets, were used to manufacture the bone artifacts recovered from the Native Alaskan Neighborhood (after findings in Binford 1978, 1981; Walker and Long 1977). The vast majority of bone tools and tool production debris are concentrated in the southern part of the Native Alaskan Village Site (Wake 1995, 1997a).

There is little ethnographic or historical information on Native American bone tool production techniques and technology. Even less has been written about Native American bone tool production techniques or technology in interethnic contact period situations, although Davydov (1977 [1810]) does briefly describe bone tool production on Kodiak Island in the late 1700s and early 1800s. It is commonly assumed that traditional stone and bone tools were quickly replaced by metal analogs in eighteenth- and nineteenth-century European fur-trading colonies and other contact situations (Kardulias 1990; Rogers 1990; Rogers and Wilson 1993; Whelan 1993). However, that appears not to be as cut-and-dried at the Ross Colony.

DIAGNOSTIC BONE ARTIFACTS FROM NAN

A wide variety of tool types and forms, as well as diverse stages of physical production, have been found in the bone artifact assemblage from the Native Alaskan Neighborhood. A number (n = 79) of complete and broken diagnostic tool forms have been identified, the majority related in some way to marine mammal hunting or fishing. Utilitarian objects such as buttons, awls, and fasteners are also well represented at NAN (Wake 1995, 1997a). A larger number (n = 146) of worked bone objects diagnostic of various technical stages of tool production have also been recovered and identified. Most of the worked bone artifacts recovered from these investigations, while clearly modified, are relatively amorphous bits and flakes of bone that defy classification as formal tools. Given the clear evidence of on-site bone tool production at NAVS, this latter category of artifacts is most likely composed of waste from various stages of manufacture or repair.

Fishing Implements

Some (n = 14) of the bone artifacts from the Native Alaskan Neighborhood are fishing implements. The most common fishing-related artifacts are barbed sections of compound fishhooks (figure 9.3a). The three relatively complete barbs and the fragment retaining the basal portion of a barb are all remains of simple unibarbs, morphologically reminiscent of fish hook barbs from Kodiak Island (Clark 1974a, 1974b, personal communication 1995; Heizer 1956; Hrdlička 1944; for illustrations and more detailed descriptions see Wake 1995, 1997a).

Utilitarian Items

A variety of bone artifacts not related to hunting or fishing have also been recovered and studied. These include broken awl tips, buttons, fasteners, crosshatched

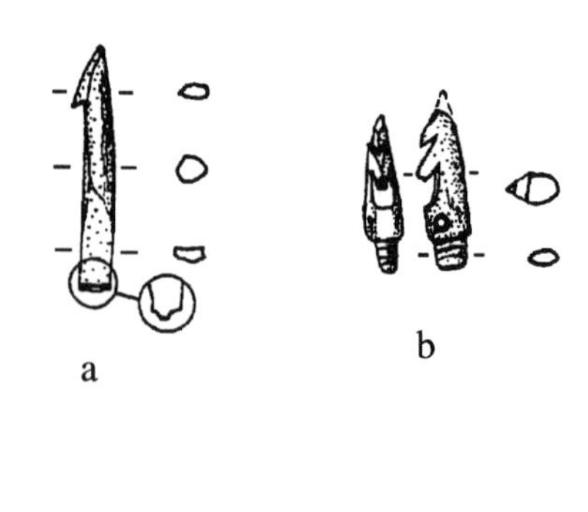

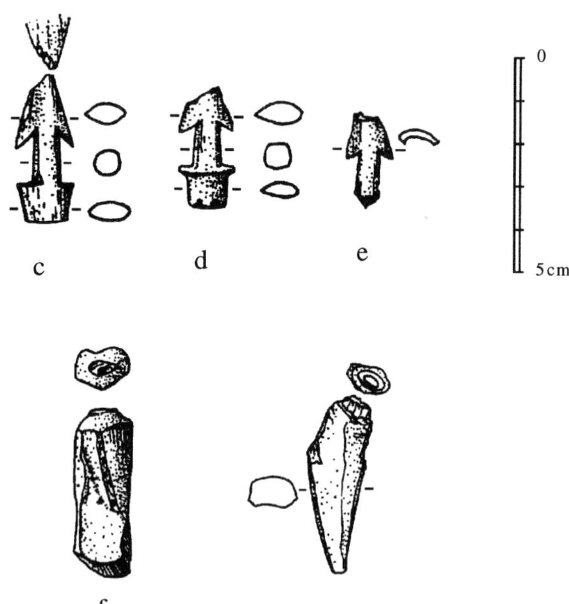

Figure 9.3. Bone artifacts from the Native Alaskan Neighborhood (reprinted with permission, from Lightfoot et al. 1997). a, fish hook barb; b, harpoon arrow point; c–e, sea otter dart points; f–g, handholds.

decorative objects, a brush fragment, a whale bone platter, and an antler baton or club (for illustrations and more detailed descriptions see Wake 1995, 1997a).

Marine Mammal Hunting Implements

Many of the diagnostic bone artifacts from Ross were used to hunt marine mammals. The marine mammal hunting assemblage consists of various projectile points and point fragments, dart socket pieces and fragments, and finger rests, most of which are specifically associated with sea otter hunting (Jochelson 1925; Ogden 1941; Scammon 1874; for detailed descriptions see Wake 1995, 1997a).

The two most common types of bone projectile points recovered from the Native Alaskan Neighborhood are arrow and dart points (figure 9.3b–e), both strongly associated with sea otter hunting (Jochelson 1925; Ogden 1941; Scammon 1874). One of these types (figure 9.3b) is a small detachable arrow point. The other diagnostic type (figure 9.3c–e) is a detachable dart point tipping a long shaft, which was typically cast from a customized throwing board similar to an atlatl. This type of dart point is a diagnostic Late Precontact style rarely found outside the Aleutian Island chain (Jochelson 1925). It is, however, present at Ross, on the Farallon Islands (where Russians had established a small outpost), and at RAC sites in the Kurile Islands—all places where Alaskan hunters were present (Khlebnikov 1976, 1990; Riddell 1955; Shubin 1990; Wake 1995, 1997a; White 1995).

NONDIAGNOSTIC WORKED BONE ARTIFACTS FROM NAN

The majority of worked bone artifacts recovered from the Native Alaskan Neighborhood at Ross are not formal, diagnostic tools or implements per se. They are, however, directly related to the on-site production, maintenance, and repair of the formal tools noted previously (and of Native Alaskan bone tool production in general). These nondiagnostic artifacts include bone and antler cores, handholds, chopped and carved waste chunks, and split bone, but in the main they consist of various chopped and carved bone flakes (Wake 1995, 1997a).

Cores

Among the nondiagnostic specimens, I have identified large chunks of raw material from which smaller pieces have been removed for further reduction or use. These large chunks evidence numerous cut and chop marks, indicative of the intensive reduction of an original skeletal element to a useful blank that will be further modified into a more formally shaped artifact. Borrowing from lithic technological nomenclature, I call these large, heavily modified chunks of bone "cores."

One particular core is identifiable as a large sectioned whale rib exhibiting evidence of numerous encircling and splitting blows delivered by metal tools (Wake 1995, 1997a). Other waste cores include basal portions of extremely large pieces of elk (*Cervus elaphus*) antler that have been thoroughly abused by metal tools during the removal of smaller bits. Two grizzly bear (*Ursus arctos*) elements show distinct signs of use as cores. The distal portions of these bones have been removed by chopping all around the circumference of the shaft of the element with a heavy-bladed metal tool such as a large knife or cleaver.

Flakes

Approximately 594 bone flakes have been recovered from NAVS, but to date none have been found down at the Fort Ross Beach Site. These flakes come in a variety of shapes and sizes, and those from Ross can be subdivided into two classes: chopping flakes and carving flakes. Chopping flakes have a wider range of sizes, indicating a lesser degree of control of individual blows than the relatively standardized carving flakes. Chopping flakes tend to be more or less square or rectangular in plan view, and generally polygonal in cross section. The variable sizes and forms of these flakes imply rapid, coarsely controlled, patterned removal of excess bone material. The characteristics of the carving flakes are somewhat different. They are generally triangular or very slim polygons in cross section and of roughly similar lengths (about 3–4 centimeters), which imply relatively finely directed and controlled force compared with chopping flakes.

Amorphous Bone Chunks

A variety of amorphous worked bone chunks and pieces are represented in the bone artifact assemblage at Ross. All these objects have indications, sometimes quite obvious, of reduction and working by metal-cutting and chopping tools. The artifacts of this category, although relatively amorphous and difficult to classify, are important data, for they further illustrate the intensity and nature of onsite bone implement production, use, and repair activities practiced in the Native Alaskan Neighborhood at Ross.

Handholds

A number of objects exhibiting various stages in the bone artifact production sequence have been recovered from both NAVS and FRBS (figure 9.3f, 9.3g); the term "handhold" best describes them. These artifacts appear to be remnant, less modified portions of the original raw bone material used by the carver. In essence, handholds probably represent the relatively unmodified portion of bone the carver held while making an implement. Items similar to these, found still attached to nearly complete bone implements, have been recovered from the Oregon coast (Lyman 1991:191, figure 5.1c).

All handholds have two primary attributes in common: narrowed, scored, cut, chopped, or snapped-off ends; and traces of cutting and carving marks indicative of more than one stage in the production of an intended tool. Some handholds exhibit as many as four stages in the tool production sequence (figure 9.4a, 9.4b) including splitting, rough carving, fine carving, and handhold removal. Handholds are the last piece of material removed from an artifact prior to its actual completion as a tool. Final smoothing and sharpening of surfaces represent the last production stage, after removal of the handhold from its source.

In general, the number of handholds present at a site is a better measure of production intensity than raw counts of finished or broken tools. Finished tools

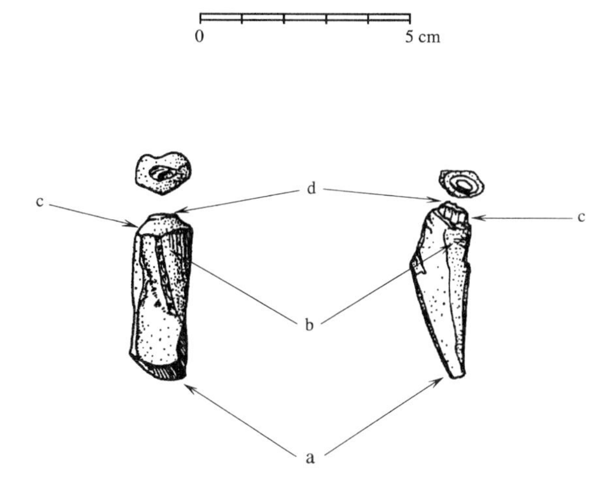

Figure 9.4. Reduction stages visible on handholds (reprinted with permission, from Lightfoot et al. 1997). a, splitting; b, rough carving; c, fine carving; d, handhold removal.

are often taken from the areas where they were made and are not brought back. In lieu of debitage, there is no sure way to determine whether excavated artifacts were actually manufactured on site. Although broken tools may be brought back to a site either for repair or still embedded in their prey, this does not say much about on-site tool production. Debitage representative of tool production and tools in their final production stages tend to remain at the place of manufacture and are the best measure of on-site tool production (for a similar argument see Dobres 1995).

BONE TOOL PRODUCTION AT ROSS

The tremendous amount of nondiagnostic worked bone material recovered from the Native Alaskan Neighborhood is testimony to the importance of the on-site production and maintenance of bone toolkits related to marine mammal hunting, fishing, and a variety of daily activities at Ross. The actual reduction sequence resulting in any given tool form probably varied somewhat, depending on specific details involved in producing the desired object, such as the addition of barbs, drilling of lashing holes, or other specific details. However, there appears to be a template of generalized stages followed in the manufacture of bone tools at Fort Ross: core preparation, core reduction, blanks, rough shaping, fine shaping, and finishing work. Core reduction and rough shaping are indicated by the great number of bone chopping flakes found in the South Trench area of the Native Alaskan Neighborhood (Wake 1995, 1997a). Subsequent fine shaping and finishing work is suggested by the recovery of carving flakes, and by various scored and snapped bits of bone (figures 9.3f, 9.3g, 9.4a, 9.4b).

Evidence suggests that scoring bone blanks with metal knives and subsequent snapping by hand was one of the primary reduction and fabrication techniques used in bone artifact production at Ross. The scoring-and-snapping technique was used in a relatively rough fashion on large objects such as whale ribs, elk antlers, and grizzly bear bones and is perhaps best seen in the handholds. Scoring, which is the encircling of a desired piece of modified or unmodified bone around its circumference with a series of cuts or chops, was accomplished using metal tools that could be controlled precisely. Most of the scored and snapped artifacts appear to have been worked with small to medium-size metal knives. Metal knife blades, somewhat smaller but reminiscent of the classic Hudson's Bay Company trading knife (Ray 1988), were recovered from the Native Alaskan Neighborhood (Lightfoot et al. 1997).

People accustomed to Western industrial craft production, or who are attempting to emulate them, would probably use metal saws to effect the same ends as the scoring-and-snapping technique, if they were available. According to the historical record (which discusses cooperage, carpentry, and shipbuilding at the colony), metal-bladed saws were available at Ross (Khlebnikov 1976, 1990). Such saws were not used to butcher animals for food consumed at Ross, however, and only rarely (three cases) is there evidence of their use in the production of bone implements (Wake 1995, 1997a).

One sectioned Steller's sea lion (*Eumatopias jubatus*) femur is of interest in this regard. On the rare occasion when metal saws were employed to fabricate bone implements at Ross, a non-European "mindset" guided their use, as the scoring-and-snapping technique was also employed. The specimen in question has been sectioned around its circumference more than once, the last such occurrence leaving a telltale bone spur indicative of the incomplete cutting of the element with a saw (Wake 1995a, figure 5.12; 1997a, figure 11.5). More important, the remaining portion of this femur has a deep groove sawed around its entire circumference. The terminations of this groove are offset and meet in a jagged manner on the ventral side. This groove, probably cut in preparation for the removal of another ring of sea lion bone, was painstakingly sawed into the cortical bone of the femur, exposing in only a few places the underlying dense, cancellous tissue.

This kind of laborious sectioning of a femur does not follow normative European or Russian practices known at the time. Numerous studies of nineteenth-century American archaeological faunal remains and European bone implement production strategies that bother to mention the use of metal saws to process mammals suggest that the vast majority of bones were cut cleanly through the entire element, typically leaving behind no telltale spurs or lips (Langenwalter 1980, 1987; Lyman 1977, 1987a, 1987b; MacGregor 1985; Shulz and Gust 1983).

At Ross, tools made of bone were clearly preferred for hunting the most economically important marine mammals, sea otters, and no metal tools for sea otter or other marine mammal hunting have been recovered from archaeological excavations. In fact, no mention of metal tool use to hunt sea otters or any other marine mammals is found in the available historical record, other than the occasional use of firearms. Although some metal harpoon points have been found in Alaska, they primarily date to after the 1850s (De Laguna 1975). Bone toolkits used by Native Alaskans were probably easier to maintain and produce than metal ones and were certainly less costly. Moreover, in terms of function and portability, techniques involved in the production of tools and implements made of bone (as opposed to metal) were just as refined and perhaps even more efficient; raw bone was also probably more readily attainable at Ross than processed metal.

DISCUSSION: INFERRING CULTURE FROM A DETAILED TECHNOLOGICAL STUDY

The evidence previously discussed suggests that the Alaskan men producing the bone tools at Ross substituted their traditional precontact manufacturing implements (which were most likely stone cutting and grinding tools) with more efficient metal-edged blades, obtained either directly or traded down-the-line, from European sources. Native Alaskans had limited knowledge of metal prior to European contact, mostly in the form of iron ship fittings washed ashore and possibly contact with some Siberian people, before European contact (Jordan and Knecht 1988; Knecht and Jordan 1985). This limited prior knowledge made metal tools highly desirable for many reasons, including their comparative rarity, but especially because metal tools hold a sharp edge longer than their stone counterparts.

Although the manufacturing implements used to fabricate bone hunting tools were different, the suite and "style" of bone tool production techniques and artifact forms recovered from the Native Alaskan Neighborhood at Ross appear to have changed little from late precontact times. One might say that though traditional production tools were replaced, and metal cutting tools were wholly incorporated into the preexisting cultural system, the mental template of sequential manufacturing techniques employed to produce bone tools at Ross remained faithful to the precontact tradition (see also Ridington chapter 8 in this volume).

This study of the worked bone assemblage recovered from the Native Alaskan Neighborhood at Ross provides information regarding two important points. First, hitherto unknown intensive bone tool production was practiced at Ross during the tenure of the Russian-American Company, between 1812 and 1841. Dependence on bone tools and related marine mammal hunting technology is, at first glance, somewhat anomalous at a Euro-Russian peri-industrial outpost, since it is commonly assumed that bone and stone tools are rapidly replaced by metal analogs. Second, the intensity of bone tool production at Ross indicates the profound importance of the use of bone tools and technology to those working and living in the Native Alaskan Neighborhood, and to the Russian-American Company's entire operation in California. Although bone tool production and use at Ross are mentioned only in passing by eyewitness observers (Khlebnikov 1976, 1990; Scammon 1874; Tikhmenev 1978), it was clearly a vital activity pursued with skill by the inhabitants of the Native Alaskan Neighborhood at Ross, as archaeological investigation has revealed.

Bone tools were important for both functional and social reasons. The intensity of bone tool production and the persistence of precontact artifact technology and styles at Ross indicate that Native Alaskan craftsmen found such tools superior to similar tools made of metal. For spiritual reasons, metal was apparently "bad" for hunting marine mammals (Knecht and Jordan 1985). Some Native Alaskans believed that metal could frighten away marine mammal prey or contaminate the success of the hunt. On the more practical side, metal is also generally denser than bone and therefore heavier per unit volume. Logically, then, the weight of metal tools could likely reduce the efficiency of a traditional style (dart or arrow) projectile delivery system, by shortening the overall range of a projectile or by throwing off a person's aim. With practice one could adapt to such differences in balance, but why bother? Moreover, metal also corrodes rapidly in salty, moist, marine environments, requiring new and different kinds of artifact maintenance.

It is clear that metal cutting implements were used almost exclusively in the manufacture of bone tools at Ross (Wake 1995, 1997a). Evidence for the use of metal implements is seen in every stage of bone tool reduction, from primary core reduction to the final finishing of projectile points. For production needs, iron tools such as knives and hatchets were more efficient and superior to their nonmetal (that is, stone) precursors: metal cutting tools hold a sharp edge longer than stone and are less easily broken during repetitive hard work. The fact that traditional Alaskan bone tool forms were still being manufactured at Ross using metal tools indicates that the craftsmen making them had experienced some degree of culture change or accommodation but were by no means fully acculturated to the Euro-Russian world view. Native Aleut and Koniag craftsmen brought with them to California their incomparable skills at sea otter hunting and incorporated European-style metal tools into their traditional tool production repertoires, which had otherwise changed little from precontact times.

CONCLUSIONS

The very act of producing bone tools may likely have meant more to Ross Colony hunters than simply ease of manufacture and functional effectiveness. By sustaining traditional bone tool production techniques, Native Aleut and Koniag craftsmen continued to identify with and display in daily practice their cultural affiliation for all to see (after Dobres and Hoffman 1994; Lechtman 1977; McGuire 1982; Weissner 1984; Wobst 1977; also Ingold Foreword and Ridington chapter 8 in this volume). These individuals were engaging in a sort of technical ritual that created and sustained core social values deemed essential to their community's viability (Pfaffenberger chapter 7 in this volume). Such activities may also indicate a kind of resistance to material domination by a colonial culture (McGuire and Paynter 1991). Though the traditional ritual of bone tool production and use was altered by incorporating metal cutting implements into the manufacturing process, the underlying themes and operational strategies and sequences remained consistent with precontact stylistic aesthetics, functionality, and cultural identity. In fact, metal tool-making implements may have been incorporated into the traditional Native Alaskan technological system of bone working at Ross as a means of reproducing maritime hunting technologies more efficiently.

The production and use of bone tools and artifacts was an integral part of the overall hunting-based economy practiced at Ross. Clearly, the colony could not have been viable, and the RAC not profitable, without the presence and traditional practices and skills of Native Alaskans and the bone tools they produced and used with remarkable efficiency. These implements and the techniques by which they were made and utilized exhibit a reification and reaffirmation of Native Alaskan cultural tradition at the colony. They also point to the importance of traditional Alaskan practices in the hunting of marine mammals critical both to the economic success of the RAC and to the maintenance of Aleut and Koniag cultural identities in California.

NOTE

I express my thanks to Chris Hoffman and Marcia-Anne Dobres for their encouragement in developing this chapter and their insightful comments. Kent Lightfoot's comments are also greatly appreciated. The artifacts were illustrated by Judith Ogden. Research resulting in this publication was funded in part by a National Science Foundation grant (SBR-9304297) to Kent Lightfoot.

REFERENCES CITED

Allan, J.

1997 Searching for California's First Shipyard: Remote Sensing Surveys at Fort Ross. *Kroeber Anthropological Society Papers* 81:50–83.

Barrett, S. A.

1952 Material Aspects of Pomo Culture, Parts 1 and 2. *Bulletin of the Public Museum of the City of Milwaukee* 20(1,2):1–508.

Binford, L. B.

1978 *Nunamuit Ethnoarchaeology.* Academic Press, New York.

1981 *Bones: Ancient Men and Modern Myths.* Academic Press, New York.

Black, L. T.

1977 The Konyag (The Inhabitants of the Island of Kodiak) by Iosaf [Bolotov] (1794–1799) and by Gideon (1804–1807). *Arctic Anthropology* 14(2):79–106.

1988 The Story of Russian America. In *Crossroads of Continents: Cultures of Siberia and Alaska,* edited by W. W. Fitzhugh and A. L. Crowell, pp. 70–88. Smithsonian Institution Press, Washington, D.C.

Chevigny, H.

1965 *Russian America: The Great Alaskan Adventure, 1741–1867.* Viking Press, New York.

Clark, D. W.

1974a *Koniag Prehistory: Archaeological Investigations at Late Prehistoric Sites on Kodiak Island, Alaska.* Tubinger Monographien zur Urgeschichte, Band 1. Verlag W. Kohlhammer, Stuttgart.

1974b *Contributions to the Later Prehistory of Kodiak Island, Alaska.* Mercury Series, Archaeological Survey No. 20. National Museum of Man, Ottawa.

Crowell, A. L.

1988 Dwellings, Settlement, and Domestic Life. In *Crossroads of Continents: Cultures of Siberia and Alaska,* edited by W. W. Fitzhugh and A. L. Crowell, pp. 194–208. Smithsonian Institution Press, Washington, D.C.

1994 *World System Archaeology at Three Saints Harbor, an 18th Century Russian Fur Trade Site on Kodiak Island.* Unpublished Ph.D. dissertation, Department of Anthropology, University of California, Berkeley.

Dauenhauer, N., and R. Dauenhauer

1990 The Battles of Sitka, 1802 and 1804, from Tlingit, Russian, and Other Points of View. In *Russia in North America: Proceedings of the 2nd International Conference on Russian America,* edited by R. A. Pierce, pp. 6–23. Limestone Press, Fairbanks, Alaska.

Davydov, G. I.

1977 [1810] *Two Voyages to Russian America, 1802–1807.* Translated by C. Bearne, and edited by R. A. Pierce. Materials for the Study of Alaskan History, No. 10. Limestone Press, Fairbanks, Alaska.

De Laguna, F.

1975 *The Archaeology of Cook Inlet, Alaska.* 2d ed. Alaska Historical Society, Anchorage.

Dmytryshyn, B., and E. A. P. Crownhart-Vaughan

1988 *To Siberia and Russian America: Three Centuries of Russian Eastward Expansion.* Vol. 2: *Russian Penetration of the North Pacific Ocean: A Documentary Record, 1700–1799.* Oregon Historical Society Press, Portland.

Dmytryshyn, B., E. A. P. Crownhart-Vaughan, and T. Vaughan

1989 *To Siberia and Russian America: Three Centuries of Russian Eastward Expansion.* Vol. 3:

The Russian American Colonies: A Documentary Record, 1798–1867. Oregon Historical Society Press, Portland.

Dobres, M-A.

1995 *Gender in the Making: Late Magdalenian Social Relations of Production in the Eastern French Pyrénées*. Ph.D. dissertation, Department of Anthropology, University of California, Berkeley. University Microfilms, Ann Arbor.

Dobres, M-A., and C. R. Hoffman

1994 Social Agency and the Dynamics of Prehistoric Technology. *Journal of Archaeological Method and Theory* 1(2):211–258.

Du Four, C. J.

1933 *The Russian Withdrawal from California*. Special Publication No. 7, pp. 52–88. California Historical Society, San Francisco.

Essig, E. O.

1933 *The Russian Settlement at Ross*. Special Publication No. 7, pp. 3–21. California Historical Society, San Francisco.

Farris, G. J.

1981 *Preliminary Report of the 1981 Excavations of the Fort Ross Fur Wearhouse*. MS on file, California Department of Parks and Recreation, Resource Protection Division, Sacramento.

1988 A French Visitor's Description of the Fort Ross Rancheria in 1839. *News from Native California* 2:22–23.

1989 The Russian Imprint on the Colonization of California. In *Columbian Consequences*. Vol. 1: *Archaeological and Historical Perspectives on the Spanish Borderlands West*, edited by D. H Thomas, pp. 481–498. Smithsonian Institution Press, Washington, D.C.

1990 Fort Ross, California: Archaeology of the Old Magazine. *In Russia in North America: Proceedings of the 2nd International Conference on Russian America*, edited by R. A. Pierce, pp. 475–505. Limestone Press, Fairbanks, Alaska.

Fedorova, S. G.

1973 *The Russian Population in Alaska and California in the Late 18th Century–1867*. Translated by R. A. Pierce and A. S. Donnelly. Limestone Press, Fairbanks, Alaska.

1975 *Ethnic Processes in America*. Occasional Papers No. 1. Anchorage History and Fine Arts Museum, Anchorage, Alaska.

Fitzhugh, W. W., and A. L. Crowell

1988 *Crossroads of Continents: Cultures of Siberia and Alaska*. Smithsonian Institution Press, Washington, D.C.

Gifford, E. W.

1940 Californian Bone Artifacts. *University of California Anthropological Records* 3(2):i–iv, 153–237.

Heizer, R. F.

1956 Archaeology of the Uyak Site, Kodiak Island, Alaska. *University of California Anthropological Records* 17(1):i–vi, 1–199.

Hoover, R. L.

1985 The Archaeology of Spanish Sites in California. In *Comparative Studies of the Archaeology of Colonialism,* edited by S. L. Dyson, pp. 93–114. BAR International Series No. 233. British Archaeological Reports, Oxford.

Hrdlička, A.

1944 *The Anthropology of Kodiak Island.* The Wistar Institute of Anatomy and Biology, Philadelphia.

Istomen, A. A.

1992 *The Indians at the Ross Settlement; According to the Censuses by Kuskov, 1820–1821.* Fort Ross Interpretive Association, Jenner, California.

Jacobs, M. Jr.

1990 Early Encounters between the Tlingit and the Russians. In *Russia in North America: Proceedings of the 2nd International Conference on Russian America,* edited by R. A. Pierce, pp. 1–5. Limestone Press, Fairbanks, Alaska.

Jackson, R. H.

1983 Intermarriage at Fort Ross: Evidence from the San Rafael Mission Baptismal Register. *Journal of California and Great Basin Anthropology* 5:240–241.

Jochelson, W. I.

1925 *Archaeological Investigations in the Aleutian Islands.* Publication No. 367. Carnegie Institution, Washington, D.C.

Jordan, R. H., and R. A. Knecht

1988 Archaeological Research on Western Kodiak Island, Alaska: The Development of Koniag Culture. In *Aurora: The Late Prehistoric Development of Alaska's Native People,* edited by R. D. Shaw, R. K. Harritt, and D. E. Dumond, pp. 225–306. Monograph Series No. 4. Alaska Anthropological Association, Anchorage.

Kardulias, P. N.

1990 Fur Production as a Specialized Activity in a World System: Indians in the North American Fur Trade. *American Indian Culture and Research Journal* 14(1):25–60.

Kari, J. M.

1990 Two Upper Ahtna Narratives of Conflict with Russians. In *Russia in North America: Proceedings of the 2nd International Conference on Russian America,* edited by R. A. Pierce, pp. 24–35. Limestone Press, Fairbanks, Alaska.

Khlebnikov, K. T.

1976 *Colonial Russian America: Kyrill T. Khlebnikov's Reports, 1817–1832.* Translated and edited by B. Dmytryshyn and E. A. P. Crownhart-Vaughan. Oregon Historical Society, Portland.

1990 *The Khlebnikov Archive. Unpublished Journal (1800–1837) and Travel Notes (1820, 1822, and 1824).* The Rasmuson Library Historical Translation Series, vol. 5. Translated by J. Bisk and edited by L. Shur. University of Alaska, Fairbanks.

Knecht, R. A., and R. H. Jordan

1985 Nunakakhnak: An Historic Period Koniag Village in Karluk, Kodiak Island, Alaska. *Arctic Anthropology* 22(2):17–35.

Langenwalter, P. E.

1980 The Archaeology of 19th Century Subsistence at the Lower China Store, Madera County, California. In *Archaeological Perspectives of Ethnicity in America,* edited by R. L. Schuyler, pp. 102–112. Baywood, New York.

1987 Mammals and Reptiles as Food and Medicine in Riverside's Chinatown. In *Wong Ho Leun: An American Chinatown.* Vol. 2: *Archaeology,* pp. 53–106. The Great Basin Foundation, San Diego.

La Place, C.

1986 [1839] Description of a Visit to an Indian Village Adjacent to Fort Ross by Cyrille LaPlace, 1839. Translated by G. J. Farris. In *Cultural Resource Survey at the Fort Ross Campground, Sonoma County, California,* edited by G. J. Farris, pp. 66–81. Report on file at the California Department of Parks and Recreation Archaeology Laboratory, West Sacramento, California.

Lechtman, H.

1977 Style in Technology: Some Early Thoughts. In *Material Culture: Styles, Organization, and Dynamics of Technology,* edited by H. Lechtman and R. S. Merrill, pp. 3–20. West Publishing Co., St. Paul, Minn.

Lightfoot, K. G., T. A. Wake, and A. M. Schiff

1991 *The Archaeology and Ethnohistory of Fort Ross, California.* Vol. 1: *Introduction.* Contributions of the University of California Archaeological Research Facility No. 49. Berkeley, California.

1993 Native Responses to the Russian Mercantile Colony of Fort Ross, Northern California. *Journal of Field Archaeology* 20:159–175.

Lightfoot, K. G., A. M. Schiff, and T. A. Wake (editors)

1997 *The Archaeology and Ethnohistory of Fort Ross, California.* Vol. 2: *The Native Alaskan Neighborhood.* Contributions of the University of California Archaeological Research Facility No. 55. Berkeley, California.

Lyman, R. L.

1977 Analysis of Historic Faunal Remains. *Historical Archaeology* 11(1):67–73.

1987a Archaeofaunas and Butchery Studies: A Taphonomic Perspective. *Advances in Archaeological Method and Theory* 10:249–337.

1987b On the Zooarchaeological Measures of Socioeconomic Position and Cost-Efficient Meat Purchases. *Historical Archaeology* 21(1):58–66.

1991 *Prehistory of the Oregon Coast: The Effects of Excavation Strategies and Assemblage Size on Archaeological Inquiry.* Academic Press, New York.

MacGregor, A. G.

1985 *Bone, Antler, Ivory, and Horn: The Technology of Skeletal Materials since the Roman Period.* Croom Helm, London.

McGuire, R. H.

1982 The Study of Ethnicity in Historical Archaeology. *Journal of Anthropological Archaeology* 1:159–178.

McGuire, R. H., and R. Paynter

1991 The Archaeology of Inequality: Material Culture, Domination, and Resistance. In *The Archaeology of Inequality,* edited by R. H. McGuire and R. Paynter, pp. 1–27. Basil Blackwell, London.

Mills, P. R.

1994 Alaskan Hunting Technologies and Cultural Accommodation at Fort Ross (1812–1841), California. *Proceedings of the Society for California Archaeology* 7:33–40.

Mills, P. R., and A. Martinez (editors)

1997 *The Archaeology of Russian Colonialism in the North and Tropical Pacific.* Kroeber Anthropological Society Papers No. 81. University of California, Berkeley.

Ogden, A.

1941 *The California Sea Otter Trade, 1784–1848.* University of California Press, Berkeley.

Okun S. B.

1951 *The Russian-American Company.* Translated by C. Ginsburg. Harvard University Press, Cambridge, Mass.

Pierce, R. A.

1988 Russian and Soviet Eskimo Indian Policies. In *History of Indian-White Relations,* edited by W. E. Washburn, pp. 119–127. Handbook of North American Indians, vol. 4, W. C. Sturtevant, general editor. Smithsonian Institution Press, Washington, D.C.

Ray, A. J.

1988 The Hudson's Bay Company and Native People. In *History of Indian-White Relations,* edited by W. E. Washburn, pp. 335–350. Handbook of North American Indians, vol. 4, W. C. Sturtevant, general editor. Smithsonian Institution Press, Washington, D.C.

Riddell, F. A.

1955 *Archaeological Excavations on the Farallon Islands, California.* Archaeological Survey Report No. 32, in Papers on California Archaeology No. 34. University of California, Berkeley.

Rogers, J. D.

1990 *Objects of Change: The Archaeology and History of Arikara Contact with Europeans.* Smithsonian Institution Press, Washington, D.C.

Rogers, J. D., and S. M. Wilson

1993 *Ethnohistory and Archaeology: Approaches to Postcontact Change in the Americas.* Plenum Press, New York.

Rousselot, J-L., W. W. Fitzhugh, and A. L. Crowell

1988 Maritime Economies of the North Pacific Rim. In *Crossroads of Continents: Cultures of Siberia and Alaska,* Edited by W. W. Fitzhugh and A. L. Crowell, pp. 151–172. Smithsonian Instutution Press, Washington, D.C.

Scammon, C.

1874 *The Marine Mammals of the Northwestern Coast of North America, Described and Illustrated: Together with an Account of the American Whale Fishery.* John H. Carmany, San Francisco.

Schulz, P. D., and S. Gust

1983 Faunal Remains and Social Status in Nineteenth Century Sacramento. *Historical Archaeology* 17(1):44–53.

Shubin, V. O.

1990 Russian Settlements in the Kurile Islands in the 18th and 19th Centuries. In *Russia in North America: Proceedings of the 2nd International Conference on Russian America,* edited by R. A. Pierce, pp. 425–450. Limestone Press, Fairbanks, Alaska.

Spencer-Hancock, D., and W. E. Pritchard

1981 Notes to the 1817 Treaty between the Russian-American Company and the Kashaya Pomo Indians. *California History* 59(4):306–313.

1982 The Chapel at Fort Ross: Its History and Reconstruction. *California History* 61(1):2–17.

Stross, F. H., and R. F. Heizer

1974 *Ethnographic Observations on the Coast Miwok and Pomo by Contre-Admiral F. P. Von Wrangell and P. Kostromitinov of the Russian Colony Ross, 1839.* Archaeological Research Facility, University of California, Berkeley.

Tikhmenev, P. A.

1978 *A History of the Russian American Company.* Translated and edited by R. A. Pierce and A. S. Donnelly. University of Washington Press, Seattle.

Treganza, A. E.

1954 *Fort Ross: A Study in Historical Archaeology.* Reports of the University of California Archaeological Survey 23:1–26.

Veltre, D. W.

1990 Perspectives on Aleut Culture Change during the Russian Period. In *Russian America: The Forgotten Frontier,* edited by B. S. Smith and R. J. Barnett, pp. 175–184. Washington State Historical Society, Tacoma, Washington.

Wake, T. A.

1994 Social Implications of Mammal Remains from Fort Ross, California. *Proceedings of the Society for California Archaeology* 7:19–32.

1995 *Mammal Remains from Fort Ross: A Study in Ethnicity and Culture Change.* Unpublished Ph.D. dissertation, Department of Anthropology, University of California, Berkeley.

1997a Bone Artifacts from the Fort Ross Beach Site (CA-SON 1878/H) and the Native Alaskan Village Site (CA-SON-1879/H), with Comments on Tool Production. In *The Archaeology and Ethnohistory of Fort Ross, California.* Vol. 2: *The Native Alaskan Neighborhood,* edited by K. G. Lightfoot, A. M. Schiff, and T. A. Wake, pp. 248–278. Contributions of the University of California Archaeological Research Facility No. 55. Berkeley, California.

1997b Mammal Remains from Colony Ross. In *The Archaeology and Ethnohistory of Fort Ross, California.* Vol. 2: *The Native Alaskan Neighborhood,* edited by K. G. Lightfoot, A. M. Schiff, and T. A. Wake, pp. 279–309. Contributions of the University of California Archaeological Research Facility No. 55. Berkeley, California.

1997c Subsistence, Ethnicity, and Vertebrate Exploitation at the Ross Colony. *Kroeber Anthropological Society Papers* 81:84–115.

Walker, P. L., and J. C. Long

1977 An Experimental Study of the Morphological Characteristics of Tool Marks. *American Antiquity* 42(4):605–616.

Whelan, M. K.

1993 Dakota Indian Economics and the Nineteenth Century Fur Trade. *Ethnohistory* 40(2):246–276.

White, P. J.

1995 *The Farallon Islands: Sentinels of the Golden Gate.* Scottwall Associates, San Francisco.

Wiessner, P. J.

1984 Reconsidering the Behavioral Basis for Style: A Case Study among the Kalahari San. *Journal of Anthropological Archaeology* 3(3):190–234.

Wobst, M. H.

1977 Stylistic Behavior and Information Exchange. In *Papers for the Director: Research Essays in Honor of James B. Griffin,* edited by C. E. Cleland, pp. 317–342. Anthropological Papers No. 61, University of Michigan Museum of Anthropology, Ann Arbor.

Wrangell, F. P. Von

1969 Russia in California, 1833: Report of Governor Wrangell. Translated by J. Gibson. *Pacific Northwest Quarterly* 60:205–215.

10. Conclusion: Making Material Culture, Making Culture Material

CHRISTOPHER R. HOFFMAN AND
MARCIA-ANNE DOBRES

STUDIES OF TECHNOLOGIES, TECHNOLOGICAL SYSTEMS, and the impact of technology on social change have a long history within sociocultural anthropology and archaeology. Archaeologists, in particular, have long maintained an explicit interest in studying technical remains and reconstructing technological processes through time and space. Over the past decade or two, scholars across the social sciences have begun to turn their attention toward a more rigorous and thorough study of material cultures—a revitalization that somewhat vindicates unwavering archaeological interest since its very inception. However, the convergence of interest in the study of technology across the humanities, social sciences, physical sciences, and engineering fields has occurred with little intellectual cross-pollination.[1]

That said, a common thread tying together these diverse fields of study is the fact that in the societies supporting them, technology plays a crucial, albeit obscured and relatively unexamined, role. As philosophers of technology have demonstrated, however, self-reflexively examining how this obscurity operates from discipline to discipline unmasks the ideology of technological determinism and myth of technological progress that has kept technology studies not just in the background, but also narrowly focused on the study of material artifacts. Their insurgent critique has added new voices to the familiar debate over the supposed benefits of technological development, still largely assumed to be both positive and accretive. Nonetheless, and despite this critique, the current cultural, economic, and political dialogue expounding the virtues of computers and the creation of "the information society" is testimony to the enduring legacy of the deterministic model of technology inherited from the early nineteenth century. Though a minority has voiced concerns about the increasing divide between those who have access to the information "superhighway" and those who do not, such issues of people, products, politics, and power are drowned out by the public outcry over pornography on the internet.

In this volume we pose an important question: How should sociocultural anthropologists and archaeologists contribute to this revitalized multidisciplinary interest in the study of technology? Stated simply, the study of technologies and techniques will never be complete without anthropological and archaeological perspectives. After all, sociocultural anthropologists study a far wider range of societies than researchers typically do in other disciplines such as history, sociology, philosophy, or engineering, while archaeologists take a much longer historical perspective than do others and study technologies with no modern counterpart. Because they are in the business of documenting, demonstrating, and understanding the incredible diversity of human cultural repertoires through time and space, sociocultural anthropologists and archaeologists are especially well situated to fill a critical gap in interdisciplinary technology studies.

While trying to pinpoint some of the unique contributions anthropologists and archaeologists can make toward a better understanding of "this culture embedded in material" (Pursell 1985:122), this concluding chapter also highlights the specific contributions the chapters in this volume make and points out some future avenues for research made possible by them. Overall, this volume has explored three topics we consider particularly important to any meaningful study of technology, past or present: technological practice, the politics of sociotechnical agency, and the role of world views in shaping techniques and end-products.

These three topics depart from mainstream sociocultural and archaeological studies, particularly those in the latter subdiscipline. Although some recent archaeological studies have begun to shed light on the practice, politics, and world views underwriting ancient technological systems, the case studies in this volume confront them as interrelated sociomaterial dynamics. Through different emphases, each tries to show explicitly how material technology is caught in everyday webs of practice, signification, agency, and normative beliefs. In addition, in an effort to situate anthropological and archaeological understandings in the context of other disciplines and to consider social developments beyond the bounds of academia, this volume looks outward across the disciplines rather than inward upon itself.

The contributors to this volume provide informative discussions and grounded case studies about the ways in which technological practice is steeped in world views, just as it is caught up in expressions of social agency and political discourse. In their general effort to reveal the less-than-tangible facets of technological systems, they converge on a number of common themes. One of the most interesting is that regardless of their particular intellectual persuasion or the specific material technology considered, all are able to highlight technology's more social "side." Focusing on these intangibles, they reveal important parallels as well as areas for further elaboration. At the same time, taken together, they build a rich tapestry of the connections linking technology to society, and material practice to social action. In different ways, these studies highlight some of the many opportunities individual technicians and work groups make for themselves to express and manipulate identities and social relationships. They also explore how symbols, metaphors, and the negotiation of political relations materialize during the most mundane of everyday technological activities. Most important—and herein lies the particular contribution of sociocultural and archaeological understandings to related disciplines—these studies demonstrate that questions of technological practice, politics, and world views are relevant to an understanding of any and all social formations, from the most egalitarian to the most hierarchical, and from the "deep past" to recent times. Equally important, these studies begin to establish an interpretive framework with which to explore and develop related analytic methodologies.

REDEFINING TECHNOLOGY

One of the recurring themes of this volume is that technology must be defined broadly. Technology is more than the materials and sequential processes by which raw materials are transformed into cultural artifacts. Certainly, any definition of technology must include material matters, but as the chapters here show, technology is also about and cannot be divorced from social relationships; knowledge, skill, and contexts of learning; and the construction, interpretation, and contestation of symbols and power. For example, both Dobres and Pfaffenberger remind us of the forgotten legacies of Mauss and Malinowski, and it would be difficult to find a more inclusive or useful definition than Mauss's concept of techniques as "total social facts."

Because definitions shape not only what researchers identify as the appropriate focus of their work but also the kinds of interpretations they think worthwhile, redefining technology is more than a semantic exercise. Thus several contributors discuss the fact that operative assumptions about what technology "is" and "does" frequently carry over into empirical research. In particular, Hoffman examines techniques of breaking, damaging, and defacing material culture in a variety of social formations, showing that artifact "life histories" do not end with the product itself. Ingold's Foreword and Ridington's chapter examine the semantic shift in Western thought, from the historical root *technē* and its association with practice, technique, art, and skill, to a more contemporary association with rational logic, the ascendance of the machine, and the "high tech" of modern society. This reliance on an instrumentalist definition of technology has shaped more than a century of technology studies in important ways—most especially by stressing artifact over artifice and product over process (Dobres 1995, forthcoming). For example, Dobres's chapter considers the implicit gender ideologies underwriting contemporary "hard-body" definitions of technology in order to understand why much of women's "soft" or "domestic" work is not considered properly technological. Along with a broader meaning for *chaîne opératoire,* her exploration of Mauss's *enchaînement organique* points out possible avenues by which to rectify the unnatural separation of *technē* from the materiality of technological practice.

Despite our intentionally open-ended definition of technology, or perhaps because of it, the contributors to this volume study technologies in rigorous and systematic ways, and none of their conclusions can be labeled "just so" stories. All make extensive use of material evidence, are explicit in their theoretical frameworks and operative assumptions, and judicious in the use of source-side analogical reasoning. The variety of approaches they employ is nearly as diverse as the variety of subject matter examined.

This diversity is also reflected in the wide range of analytic approaches taken here: from Roux and Matarasso's empirical and multiscalar study to Pfaffenberger's interpretation of worlds of intersubjective meaning. What all the chapters share is a methodology grounded in the use of multiple sources of intellectual inspiration and multiple lines of material and social evidence. As the preceding chapters show, a holistic approach is almost certainly a prerequisite for studies whose goal is to understand the technological dimensions of culture simultaneous with the cultural dimensions of technology.

Among the analytic tools and interpretive lenses combined innovatively in this volume are *chaîne opératoire,* with its focus on technical sequences, and the "life histories" of artifacts; detailed materials analysis; an explicit concern with change and innovation grounded in historically specific sociopolitical conditions; research and interpretation at multiple phenomenological and physical scales; and subject-centered approaches able to humanize the technologies in question by focusing on the *artifice* of artifact production and use.

The volume is also explicitly concerned with developing innovative ways to identify the social forces "at work" in technology through detailed materials analysis. The volume offers a number of useful concepts and heuristic devices in this regard, all dedicated to developing a better appreciation of the social dynamics of technological practice, politics, and world views.

TECHNOLOGY EMBEDDED IN THE MATRIX OF CULTURE

One particularly important concept is that technologies, technological systems, and material techniques of artifact production and use are embedded deeply in the very matrix of culture. Rather than acting as a "subsystem" interacting at only some points along a linear continuum with others (such as subsistence or exchange), technology and culture are but two dimensions of a single highly faceted phenomenon. This is an operative premise taken up by most contributors, such that technical actions are seamlessly interwoven into the social and material lives of cultural agents, who are themselves situated within larger temporal, spatial, historical, material, and social fields. This embeddedness is an ongoing and dynamic process through which both tangible and intangible dimensions of technology work reflexively to create a cultural matrix that turns back on itself, over and over again.

As several chapters demonstrate, social identities and interpersonal relations are expressed and enacted within the arena of technological practice. These can be core identities based on economic and political status, ethnicity, gender, age, or kin affiliation; and identities can be more closely tied to a particular technology or technological practice, such as beadmaking, iron-smithing, seal hunting, *bwayma* storage hutmaking or canoe building, or domestic service. As documented here, first-order social identities can give rise to (or at the very least reaffirm) others based on the performance of technical skill or the display of technical knowledge: apprentice, weaver, elder smith, chief, master. The general view promoted here is that a person's multiple and overlapping identities are constructed in large measure through technological practice, and Larick's study of concomitant temporal changes in colonial New England house configuration and social relations of production against the backdrop of developing industrialism is a case in point. As well, Childs discusses the roles, identities, and social relationships enacted during the various stages of Toro ironworking, commencing with the search for a suitable source of iron ore through to the final stages of manufacture and proscribed redistribution of smelted products. Rules of participation in the many material and ritual stages of iron production are most frequently drawn along the lines of gender or kinship. In so linking material action to social rights and duties, Toro ironworking simultaneously creates and reinforces status-related obligations and opportunities. For Dobres, because technological gestures are typically witnessed by others in the community, they help express and manipulate salient identities while promoting (or thwarting) self-interests. As she argues with particular reference to communally organized societies, and as Pfaffenberger shows for swidden agriculturalists, "performative" aspects of technology can become a source by which to gain (or lose) status and prestige. Similarly, through detailed analysis of archaeological materials, Wake demonstrates how technological gestures can also promote ethnic or cultural identity. At the Russian outpost of Fort Ross (California), a European tool—the metal saw—was used in a distinctively non-European manner, appropriated by Native Alaskans, and thus "made" Aleutian. Identities "manufactured" through technological practice are not static entities. Rather, they are constructed, reconstructed, and often contested on an ongoing basis.

As our introduction noted, in the last two decades we have come to understand that technologies are implicated in both expressing and reaffirming normative belief systems, ideologies, and world views. These less than tangible aspects of technology, where technique and practice blur into ritual and weave themselves into the very fabric of myth and cosmological understandings of the spirit world, are particularly evident in several contributions to this volume. For example, Childs highlights the role of the spirit medium and of spirit elders in the material *chaîne opératoire* of Toro ironworking. Similarly, by quoting numerous allusions to hunting techniques and the materials employed in such activities, Ridington shows how Athapaskan cosmological narratives blur the artificial boundary between narrative, discourse, and technology; and, Dobres discusses how both underlying ideologies and everyday practices together serve to ideologically reconfigure collective subsistence efforts into supposedly male-only hunting activities.

From the ritual and cosmological dimensions of technologies, it is but another small step to cultural metaphors invoked by and (re)created during technological endeavors. As Pfaffenberger's study illustrates, salient metaphors resonate during the mundane construction and use of Trobriand *bwaymas* and Gawan canoes. These metaphors, of lightness, darkness, and heaviness, are reaffirmed over and over again in the course of everyday technological practice, further solidifying their importance to Trobriand society. Hoffman classifies acts of intentional damage as either metaphorically constructive or deconstructive depending, for example, on whether the act is intended as normative or critical of general values, and *who* is doing the damage. In his archaeological study, Hoffman then shows that patterns of intentional artifact breakage resonated with, and thus fed back on, other stages in the use-life of metal artifacts. And in Childs' study of Toro ironworking, communal metaphors of reproduction and childbirth are continually invoked and materialized during iron-smelting activities. As Dobres argues more generally, reproduction and childbirth are transformative processes metaphorically linked to the production of material culture. Finally, in her Afterword to this volume, Lechtman demonstrates that during technological performance, people employ and create ethnocategories—meaningful attitudes and understandings of materials and technologies—that are as culturally salient as those rendered linguistically.

TECHNOLOGICAL AGENCY, SOCIAL ACTION, AND PRACTICE

Demonstrating that technologies are deeply embedded in the very matrix of culture is one thing, but understanding *how* this comes about is a more significant, albeit difficult, task. The contributors to this volume demonstrate compellingly that this "embedding" of culture in technology is an ongoing dynamic process—that is, a verb of action. Technological practices are socially engendered activities during which people construct more than material objects. Through such practices they reflexively construct, reconstruct, and simultaneously reinterpret culture itself (Dobres and Hoffman 1994). This recursive view emphasizes the action and agency of technology *as* social practice. Explicit considerations of agency and the application of practice theory to an understanding of technology, while relatively new to archaeology, are variously taken up in this volume by Dobres, Hoffman, Pfaffenberger, and Ridington.

One important dimension of technological practice is the degree of heterogeneity and structural complexity underwriting the dynamics of technological culture. Within the arena of technical activities, people perceive and act on their material world as both individuals and members of collectivities who possess unique life histories, motivations, interests, and affiliations. The interesting dialectic here is that individual technicians are simultaneously members of larger communities, which act with motivations and agendas that are possibly different from their own. No doubt this multiscalar structural complexity manifests itself differently in different kinds of societies, but in all societies, heterogeneity at the level of the individual serves to shape social and material/technical actions. Furthermore, just as it is likely (perhaps even certain) that aggregates of individuals do not necessarily work toward identical goals, it is equally certain that individual participants enter into communal technological activities possessing and expressing different levels of knowledge, experience, and skill.

Thus, in this volume, technological agency is made especially visible in case studies highlighting the ongoing construction, negotiation, and contestation of social identities and social relations of production during technical activities. Such basic cultural phenomena as these are not static; personhood and social relationships are made and remade, and technology is essential to that transformative process (Dobres forthcoming). Four contributions (Childs, Hoffman, Dobres, and Pfaffenberger) remind us that during their lifetimes, individuals, much like artifacts, are transformed, one stage at a time. As individuals enter into new and different sorts of interpersonal relationships and take on different technical roles and responsibilities, they augment or outright accrue new identities. Hoffman discusses how gendered identities of manhood (among teens) are developed during cooperative efforts of destruction: vandalism and graffiti. For Childs and Pfaffenberger, social relationships and roles enacted during technical rituals entail obligations to be met later on; while for Dobres, first-order social identities can be challenged, confirmed, or "built upon" during technical activities of various kinds. And on a wholly different scale, that of generations, the dialectic of changing social relations and architectural change lies at the structural core of colonial New England house configuration, as Larick shows.

Similarly, at the same time that technological practice relies on and reinforces cultural metaphors, those meanings and metaphors can be remade, reinterpreted, and transformed during material production and use activities. For example, Pfaffenberger focuses on the reaffirmation of normative beliefs during activities as diverse as building yam storehouses and canoes and coal mining, and Hoffman highlights the multiple meanings and contestation of normative values during acts of intentional damage. Moreover, the act of writing graffiti on a subway wall has different meanings for different people; perhaps the graffiti writer is aware of some of these heterogeneous meanings, but even he or she is not omniscient.

Another component of social agency made explicit in these chapters is the role of technological choices, opportunities, and alternatives in the decision-making process of artifact production and use. During technical activities people make choices of all sorts. They are perhaps only rarely constrained to a single operational path by virtue of artifact physics. For example, Larick shows the historically configured technological field of alternatives from which colonial New Englanders drew and redrew in order to build and configure their houses over time. And, in their effort to model precisely the technical and labor requirements of Harappan bead production, Roux and Matarasso argue the importance of identifying all possible alternatives available to technicians at each stage of a productive *chaîne opératoire.* In selecting from a field of alternatives, technicians are influenced by numerous factors—material, social, and symbolic—and an individual's sensibilities and aesthetics are as much contingent factors structuring practical decisions as the objective nature of material resources. Nonetheless, it is clear that opportunities for social maneuvering and the manipulation of cultural symbols arise during every stage of artifact production and use. These opportunities very likely have direct bearing on decision-making processes at the scale of individual or group agency; we believe it is time to investigate these structuring dynamics more explicitly.

Two contributions (Ridington and Wake) consider the adoption of a foreign material—in both cases iron—into a technocultural context where it was previously unknown. Although in the Standard View of progressive technological evolution this would be considered the "natural" outcome of the introduction of a superior material into an indigenous context, recent advances in innovation studies (for example, van der Leeuw and Torrence 1989) compel us to question this assumption. While using significantly different analytic approaches, both Ridington and Wake demonstrate the complexity of the culture-contact process by focusing on socioeconomic, historic, cultural and cosmological factors. In the narrations retold to Ridington, we learn that foreign artifacts and technologies were "dreamed into" an existing oral tradition, that is, reinvented anew. These two studies strikingly demonstrate that, although the adoption of iron supposedly represents a significant technological and material "advance," continuity and tradition are much more pronounced. The prevailing technological style and world view underwriting both Alaskan and Athapaskan technologies remained conservative—in spite of material changes in the resources utilized.

Technologies carry meaning and "build" culture on occasions other than those that might be considered ritualistic or explicitly communicative. As chapters in this volume show, practice theory provides important links for understanding social action at the scale of day-to-day material activities. Clearly, such "quotidian" or seemingly mundane activities are important arenas in which the construction and reconstruction of both technical agents and culture take place. The daily sorts of activities in which people engage at this scale form a nexus of sociotechnical structures wherein motivations, metaphors, and meanings converge to help individuals determine their next course of action. As Dobres suggests, Bourdieu's concept of *habitus* is especially useful in connecting the "everyday" to the macroscale social processes, such as those considered by Roux and Matarasso, Wake, and Hoffman. In Larick's study, both interpersonal household-level dynamics and the physical architecture of colonial New England homes were situated within streams of antecedent historical and economic conditions. The daily practice of household service both reaffirmed and gave material shape to those background conditions as they simultaneously prefigured longer-term macroscalar economic and material change. Although the technological rituals discussed by Pfaffenberger are particularly illuminating examples of normative culture "in action," the worlds of intersubjective meaning forged while building canoes and constructing *bwaymas* equally rely on mundane and routinized activities for their reaffirmation. Finally, Ridington shows us that culture is "made" material during everyday activities in the hunting cultures of northern Canada. Among Athapaskans, because of the high level of shared knowledge—about mundane hunting matters as well as their cosmological underpinnings—everyday language and action are sufficient to construct a strongly normative technological tradition without appeal to overt expressions of solidarity.

Not surprisingly, one of the particularly useful concepts to emerge from this volume is the *performative* nature of technology acts. Emphasizing artifice alongside the artifact, Ridington shows how the everyday performance of Athapaskan hunting techniques is embedded in an oral narrative tradition that implicates the community at large. Similarly, the meanings of Trobriand storage and canoe technologies explored by Pfaffenberger (and first noted by Malinowski) are of a decidedly performative nature. In their material efforts to express personal interests and create social identities, Dobres's technical agents enact codes of silent discourse via performative/technical gestures. And, with reference to archaeological examples, Wake and Hoffman demonstrate how the very act of making and breaking is a powerful communicative event. In a modern context, Hoffman also reminds us that performative acts of intentional destruction are dramatic, even surprising, events.

TECHNOLOGICAL POLITICS AND POLITICAL TECHNOLOGIES

When considering the social agency and performative nature of technological practice, its political dimensions come almost unavoidably to the fore. Whether the maintenance and manipulation of political structures take place in an overt or tacit manner, technologies are constructed within existing frameworks of sociopolitical relations. Technological practice also provides a context in which acts of resistance, contestation, and, quite possibly, change may take root. Running throughout the volume and having significant potential for further exploration is the theme of technology as a political dynamic and how the production and possibly differential control of technical knowledge, techniques, resources, or end-products relate to social status.

Certainly, one question that is simultaneously technological and political (and which has long concerned archaeologists, in particular), is: What role does craft specialization play in the emergence of complex societies? Roux and Matarasso argue by example that the on-the-ground economic and material organization of craft enterprises must be examined in detail before their impact "on" society can be assessed. Their example, specific to ancient Harappan beadmaking, focuses on how we can delimit the labor requirements and organization of craft industry through *activity analysis*. Through detailed materials analysis of the *chaîne opératoire* of contemporary beadmaking, they are better able to estimate ancient labor requirements. Their conclusion suggests that in Harappan society, the material craft of beadmaking could *not* have had a significant impact "on" social developments. Rather, their findings suggest that social developments more likely led to technological (material) changes that, over time, were able to support an emergent class of craft specialists. Simply put, too few people were involved in Harappan bead production to "necessitate" the development of a complex organization of elite managers.

In several contributions, technical activities and meanings are overtly manipulated by a dominant group, either to maintain or strengthen an existing base of power or privilege. Childs documents how, during the sequential activities of Toro ironworking, elder smiths display a distinct interest in accumulating status and wealth within socially accepted bounds. But at the same time that technological practice is implicated in acquiring and maintaining relations of power, it affords opportunities for subordinate groups to critique, undermine, or usurp them. Thus, under certain conditions, technology may provide an arena for resistance and change. Wake describes how the Russian-American Company grew wealthy on the sale of otter pelts by exploiting the well-honed traditional hunting skills of Native Alaskans. At the same time, however, Native Alaskan technological expertise afforded Aleutians a means of resisting wholesale acculturation, thereby enabling them to maintain a significant degree of cultural and ethnic individuality. Childs' study shows that, although elder smiths traditionally have the upper hand in accumulating wealth and status, younger men take advantage of opportunities for personal enrichment by discovering new sources of iron or constructing new furnaces. Similarly, by participating in dances associated with certain ironworking rituals and allowing herself to be "climbed" by a spirit, an unmarried Toro girl might someday become a *nyakatagara* (or spirit medium), a position of status and power. Finally, Hoffman shows that technical activities associated with vandalism and graffiti in modern U.S. cities provide opportunities for disenfranchised youths to criticize dominant power structures. More generally, he argues that during acts of destruction, the power of made things can be manipulated in order to make statements and even usurp the symbols, the technical knowledge, or the end-product itself.

The politics of technology are evident not only in the overt examples discussed above, where meanings, relationships, identities, and economic structures are invoked or questioned through technological practice. Dobres and Ridington also point out that the performance of technical knowledge and skill can provide the means for acquiring social power and economic status. In communal societies, knowledge and skill typically serve as foundations on which privilege is built. Dobres further suggests that "mere" technical gestures, under certain performative conditions, can serve as opportunities for individuals and groups to express and augment their reputations, and thus their identities. She suggests that, at certain Upper Palaeolithic occupation sites in the Paris Basin and French Pyrénées, differential access to and the display of technical *connaissance* and skill in lithic and organic artifact manufacture and use may have served as a material means of manufacturing status and prestige over the course of one's life. And, as Ridington cogently demonstrates for contemporary hunting societies, technological knowledge held "in the mind" can be more practical and valued, by the community at large than physical tools carried about in the hand. Knowledge shared within a community is often key to its survival, and that knowledge is frequently technological. His point is that rather than the number of material technounits in their repertoire, it is communal knowledge about technical artifice that makes these societies successful and able to adapt to changing circumstances.

THE MATERIALITY OF SOCIAL AND TECHNOLOGICAL PRACTICE

In the Foreword, Ingold notes that "the perspectives offered here are as diverse as they are conflicting." Indeed, each author considers the social dynamics of technology with their own set of concerns and theoretical and methodological perspectives specific to his or her disciplinary history and to the material technology in question. What they share, nonetheless, is a desire to bring technology "back to life by reinserting it into the current of human activity and social relations." To accomplish this, the contributors converge on three salient topics: that technological practice is social relationships, that technology plays a role in the simultaneous maintenance and critique of political structures, and that cultural world views are embedded in the very structure of material technologies. The contours of such understandings are only now developing, but taken together, these studies provide a significant contribution to further articulating the shape they may take.

There is much still to understand, not only about the several concerns addressed in this volume, but especially about the materiality of technological practice. For instance, neither archaeologists nor sociocultural anthropologists have explored adequately the learning process associated with the cultural transmission of skilled technical knowledge (but see Lave and Wenger 1991). Similarly, although tactile sensory experience is clearly important to the ways humans are simultaneously material and cultural, its links to cognition and technical practice are poorly understood. Along related lines, though not necessarily seen as determinants, physical materials are traditionally treated as constraining rather than also possibly *enabling* social action. As Ingold's Foreword discusses, most approaches to technology are conceptually constrained by a distinction between mechanisms and semiotics, between action and meaning, and between what something *does* and what something *means*. He counters these distinctions with a phenomenological approach in which meaning and knowledge are "generated in, rather than expressed by, the practices in which people are engaged." In their different attempts, this is precisely what the chapters in this volume have tried to understand.

Several contributors comment on the difficulty of resolving the important questions raised in this volume, particularly in terms of generating trustworthy material (that is, testable) evidence. This problem is particularly worrisome for archaeologists who cannot directly observe the intangibles of ancient technological practice as can ethnographers and ethnoarchaeologists. But to turn the tables, neither can ethnographers and ethnoarchaeologists continue to ignore the materiality of the intangible dynamics they can "see." These case studies demonstrate conclusively that sociocultural anthropologists and archaeologists have only begun to explore the potential of intradisciplinary dialogue. Yet sociocultural anthropologists and archaeologists are especially well situated to make a significant contribution to the revitalized interdisciplinary interest in technology currently cross-cutting the humanities, social sciences, and engineering and materials sciences. We should not expect this contribution to be narrow in scope, nor will it be simple to define in a way that we can all agree on. No doubt, anthropology's and archaeology's combined contribution to a purposefully broad understanding of the social dynamics of technological practice, politics, and world views will be as diverse as the cultures and technologies it studies and as varied as the many interpretive lenses employed. But we hold that such diversity will provide a rich field from which all of us interested in the social dimensions of technology, and the technological dimensions of society, can draw inspiration.

NOTE

1. For History, see Staudenmaier (1985). For Philosophy, see Heidegger (1977), Ingold (1988), and Mitcham (1980, 1994). For Sociology, see Callon (1987), MacKenzie and Wajcman (1985), and Pinch and Bijker (1987).

REFERENCES CITED

Callon, M.

1987 Society in the Making: The Study of Technology as a Tool for Sociological Analysis. In *The Social Construction of Technological Systems,* edited by W. E. Bijker, T. P. Hughes, and T. J. Pinch, pp. 83–106. MIT Press, Cambridge, Mass.

Dobres, M-A.

1995 Of Paradigms Lost and Found: Archaeology and Prehistoric Technology, Sleepwalking through the Past. Paper presented at the Annual Meetings of the Society for American Archaeology, Minneapolis, Minn.

Forthcoming *Technology and Social Agency: Outlining an Anthropological Framework for Archaeology.* Blackwell, Oxford.

Dobres, M-A., and C. R. Hoffman

1994 Social Agency and the Dynamics of Prehistoric Technology. *Journal of Archaeological Method and Theory* 1(3):211–258.

Heidegger, M.

1977 *The Question Concerning Technology.* Translated by W. Lovitt. Garland, New York.

Ingold, T.

1988 Tools, Minds, and Machines: An Excursion in the Philosophy of Technology. *Techniques et Culture* 12:151–176.

Lave, J., and E. Wenger

1991 *Situated Learning: Legitimate Peripheral Participation.* Cambridge University Press, New York.

MacKenzie, D., and J. Wajcman (editors)

1985 *The Social Shaping of Technology: How the Refrigerator Got Its Hum.* Open University Press, Milton Keynes, England.

Mitcham, C.

1980 The Philosophy of Technology. In *A Guide to the Culture of Science, Technology, and Medicine,* edited by P. T. Durbin, pp. 282–363. Free Press, New York.

1994 *Thinking Through Technology: The Path between Engineering and Philosophy.* University of Chicago Press, Chicago.

Pinch, T. J., and W. E. Bijker

1987 The Social Construction of Facts and Artifacts. In *The Social Construction of Technological Systems,* edited by W. E. Bijker, T. P. Hughes, and T. J. Pinch, pp. 17–50. MIT Press, Cambridge, Mass.

Pursell, C. W. Jr.

1985 The History of Technology and the Study of Material Culture. In *Material Culture: A Research Guide,* edited by T. J. Schlereth, pp. 113–126. University Press of Kansas, Lawrence.

Staudenmaier, J. M.

1985 *Technology's Storytellers: Reweaving the Human Fabric.* MIT Press, Cambridge, Mass.

van der Leeuw, S. E., and R. Torrence (editors)

1989 *What's New? A Closer Look at the Process of Innovation.* Unwin Hyman, London.

Afterword

HEATHER LECHTMAN

IN HIS FOREWORD TO THIS VOLUME, Tim Ingold challenges us to specify the claims we make in using the term *technology* and to justify those claims. Here is my claim. Ethnocategories by which people order experience are rendered through technological behavior just as they are rendered linguistically (Lechtman 1996a, 1996b). Appropriate study of the materials technologies that people used in producing their objects will allow archaeologists to identify these categories, investigate them, and understand their utility as organizing principles in other spheres of social activity.

People understand and manage the physical and social world in which they live by creating ethnocategories of things, events, behaviors, and relationships that help render the world intelligible. Systems of classification are expressed linguistically, and we are comfortable in studying language to elicit the ethnocategories that serve as keys to the architecture of culture. Ethnocategories are also made behaviorally, that is, in action. Fortunately for archaeology, they are manifest in the materials technologies that were used to produce the artifacts we study. Manufacturing an object always involves accommodation between the properties of the material from which the object is made and the object's design: the possibilities and constraints any material presents in handling versus how we want the material to perform. The fact that the physical properties of natural materials are immutable and invariant wherever they are found means that variations in the ways culture-bound practitioners manage these materials reflect cultural choices. Our ability to identify cultural decisions and choices in the technologies behind object production lies precisely in this regularity in the physics of matter (Hosler 1994; Lechtman 1977; Smith 1975).

My claim is that we may identify in materials management certain cultural principles that people used to order and structure reality through technological performance, just as they organized and systematized the world through language. In saying this, there is no claim that linguistic and technological ethnocategories always coincide. Sometimes they do not. But both exist, and it is up to archaeologists to identify ethnocategories that are made concrete in artifacts by virtue of the artifacts' production histories.

ETHNOCATEGORIES AND PRINCIPLES ACCOMPANYING THE MANAGEMENT OF METAL IN THE CENTRAL ANDEAN AREA

As an example of this approach, I offer insights based on a large body of laboratory data that identify central Andean metallurgical practice in prehistory. One of the primary ways Andean objects carried and conveyed meaning was through the materials and procedures used in their manufacture (Lechtman 1977, 1984, 1993). Meaning inhered in the activity and performance of production, that is, in process as well as in product. Cultural and material attributes were realized through, and were at one with, appropriate technological performance (Lechtman 1996a).

What ethnocategories and attitudes about metal as a material can we discern from Andean metal technologies?

Metal Is a Solid Material

Andean peoples chose to focus on and to utilize those mechanical properties of metal that allow it to be shaped as a solid material: plasticity, malleability, hardening through deformation, and softening through moderate annealing. The alternative of shaping metal as a liquid material by casting it into a mould—techniques highly developed in the lost-wax castings of Colombia, Central America, and southern Mexico—was paid scant attention by Andean smiths. The distinction between these two approaches to the handling of metal is significant. The art of casting, or shaping metal as a liquid, lies in the design and preparation of the mould, not in the pouring of molten metal. The material critical to success is the refractory of which the mould is constructed. The integrity of the casting that solidifies inside is a function primarily of proper mould design. It is the mould that determines the object's shape, not the liquid metal.

Metal Is a Plastic Material

Shaping metal as a *solid* relies entirely upon the mechanical properties of the metal, most particularly on its plastic behavior. Metal deforms plastically; it alters shape under the influence of an external force, such as a hammer blow and, when the force is removed, maintains the new configuration. Andean smiths were expert in the plastic deformation of the metals and alloys they produced; they concentrated on plasticity as one of metal's most valuable mechanical properties and pushed that property to its limits (Lechtman 1988). It is clear that metal in the form of sheet, hammered to uniform thickness and at times to the thinness of foil, was highly valued in and of itself (Hosler et al. 1990).

Flat, two-dimensional sheet served also and almost universally as stock from which to build three-dimensional forms. Smiths carefully hammered and preshaped sheet metal parts, then assembled and joined them mechanically or metallurgically to construct animal and human figurines and other hollow, closed forms. No sculpture was too small for this sheet assembly treatment. Some of the most complex and imposing constructions we have in metal are miniatures that measure only 1–2 centimeters in height (Lechtman 1988).

A preponderance of Andean symbolic objects in metal appear to be of gold and silver; they are made of sheet metal. A key requirement of any sheet metal tradition is the production of sheet from material rigid enough to maintain its form. By adding copper to silver and copper to gold, Andean smiths produced alloys that performed admirably. Copper as an alloying element strengthens but also toughens silver and gold. Toughness is the opposite of brittleness; a tough metal is one that resists fracture. Binary silver-copper and gold-copper (or ternary gold-copper-silver) alloys were Andean solutions to the problem of producing metals malleable enough to perform plastically when hammered into sheet, and rigid enough to maintain the shape their very plasticity enabled them to achieve.

With these data, we can begin to contemplate some of the attitudes Andean smiths held about metal. First and foremost, metal was a solid whose mass could be and commonly was extended into thin, flat forms. Uniformity of thickness was achieved regularly, and thinness had value. The mechanical properties metalworkers sought and developed in alloys were plasticity, malleability, and toughness—not hardness, strength, and sharpness, which are properties we associate with the development of metallurgy in the ancient Old World. Plasticity, malleability, and toughness are mechanical properties of natural materials. *Planarity,* a quality that describes the state of the material from which most constructs in metal were built, is a property of Andean technological style in metal and in the management of other materials, such as cloth (Lechtman 1996a).

Alloys Are Media of Transformation

Andean metallurgy was a three-component system. The elemental or material components are copper, silver, and gold. The system was set in place as soon as these three metals were identified and used commonly, early in the Early Intermediate Period (ca. 200 B.C.–A.D. 600), and the triad remained a physical and cultural reality through the Late Horizon (ca. A.D. 1476–1532).

The most important physical property of Andean metals and alloys was their color. We recognize Andean metallic colors not primarily on the basis of linguistic evidence about color categories (Money 1998), but from consideration of the archaeological and laboratory-technical data on how metals were made and used to achieve culturally appropriate color. Metal objects, especially those made from sheet, often underwent dramatic color change during the fabrication process. Laboratory studies have shown that copper is always the medium for such color transformation (Lechtman 1973, 1984). Copper was the "mother" of Andean metals in the sense that it generated the properties Andean peoples sought in metal. Copper was the source of those properties, and the instrument of transformation.

Color was managed primarily through the development and use of two key alloy systems: the alloys of copper with silver and copper with gold, the latter commonly known as *tumbaga*. The mechanisms responsible for the dramatic color alterations these alloys undergo as they are hammered into sheet are often referred to as depletion and enrichment. The terms *depletion silvering* and *depletion gilding* describe these metallurgical color-development techniques. Copper, the necessary agent for color transformation, is always a major constituent of the alloy. The binary and ternary *tumbaga* alloys enabled objects to present culturally required metals and colors—varied hues of copper, silver, and gold—produced through transformations within the structure of the alloys themselves. Thus color, the emblem of the object, comes from inside it. In the Andes, color was the external and enhanced consequence of a change in internal state or structural order (Lechtman 1996a).

PRECISE KNOWLEDGE AND ITS CULTURAL ORGANIZATION

I have identified two components of a central Andean technological style to which prehistoric metalworkers were committed. The features of one component include planarity (two-dimensionality) and the joining of pliable parts. The other component is characterized by the development of a layered structure within the material itself and the realization of surface color through the manipulation of such structural layers. Both components were achieved through management of the metal copper.

There are some clear levels of correspondence between managing metal and managing cotton and camelid fibers in the Andean zone (Lechtman 1988, 1993, 1996a). Both resulted in the manufacture of essentially planar expanses of pliable material: metal sheet and woven web. These then served as the basic products for elaborating other objects, such as metal sculpture and cloth garments. If, however, we consider *dimensionality* rather than form as the focus of Andean manufactures in cloth and in metal, there are three dimensions to a woven web or a metal sheet: two are given by its two areal dimensions, and a third vector defines what lies inside or beneath. This third dimension corresponds to the essence of the material or the object, and is structure dependent. In structural weaving, for example, designs and color areas, that is, all the message-bearing features of the cloth, are generated by the manipulation of planes of warp and weft yarns, the structural building blocks of the woven web (Lechtman 1984, 1993, 1996a). The meaning of woven cloth in the Andes is synonymous with its rendering; structure and essence are one (Conklin 1996).

The same argument holds for the depletion and enrichment phenomena that develop culturally appropriate colors on the surface of metal objects. In analogy with woven materials, the internal structure of many worked, *tumbaga*-type alloys consists of intermeshed layers of metallic phases. The manipulation of this microstructure is what generates and releases the color.

If our aim is to identify certain aspects of the culture of a prehistoric people from their technological experience, then we need to ask the question: Do these features of Andean technological behavior—ethnocategories rendered technologically—reflect or embody, or did they even help generate, a more broadly held Andean conceptual system? Ethnohistorians and ethnographers of present-day Andean communities discuss long-held Andean beliefs in the presence of a life force or "animating essence" in all things (Allen 1988; Carpenter 1992), including manufactured objects (Harrison 1982; Taylor 1974–1976). The ethnographies also describe an Andean cosmic vision of a circulatory world (Allen 1988; Urton 1981) in which spatial and temporal components are inseparable. That world has a "subterranean interior that contains both past and future" (Allen 1988:226); the interior world incorporates events that have already occurred and that may reoccur, to emerge outside. These events, like the setting and the rising sun or the flow of celestial and terrestrial waters, pass through a plane that separates celestial space from the underworld, in a circulatory round. The earth is this plane, a mirror that alternately reflects celestial and subterranean order across and through itself (Urton 1981).

The technological ethnocategories considered in this discussion of Andean metal artifacts—solidness, plasticity, planarity, transformation—join space, or the material aspect of the world, with time, both history and the future, in that, as a system, they differentiate essences from interior states or conditions. The development of the essential qualities manifested by cloth or metal are historical processes; the end result comes from altering a previous condition and transforming it in present time. This set of ethnocategories is observable in the archaeological record as features of what people decided to do as they managed aspects of their material world. We identify them because they remain as physical facts of technological behavior. What they show is that the relations between technological performance and the shared cultural expectations that render the world intelligible to Andean peoples seem close (Lechtman 1996a).

ETHNOCATEGORIES LINGUISTICALLY AND TECHNOLOGICALLY RENDERED

The highland mining district of Julcani, in Peru's Department of Huancavelica, draws miners largely from communities in Angaraes Province, where the mine is located. They work a polymetallic silver-lead-zinc deposit, though enargite (copper sulfarsenide) ores are also present. The miners share a set of beliefs about the underground world from which they extract metallic ores and about the relationships of the ores to that world and to one another (Salazar-Soler 1992). These beliefs integrate ideas with strong roots in the pre-Hispanic past and ideas introduced by Europeans in the sixteenth and seventeenth centuries.

Ethnohistoric accounts describe certain attitudes and practices surrounding mines and the mining of metallic ores that were current among Andean communities during the Inka hegemony (Berthelot 1978). Miners selected, safeguarded, and venerated certain exceptional ore specimens they extracted from individual mines. These metallic minerals were called *mamas. Mamas* were distinguished by some special quality or suite of qualities: richness of the metal content of the ore, extraordinary size, vivid color, unusual shape, rare beauty. *Mamas* were *huacas,* sacred stones; each mine had its own *huaca,* regardless of the kind of metallic ore it represented: gold, silver, copper, mercury.

In a more general sense, we learn from early Spanish accounts that Andean peoples considered metallic minerals to be products of the earth, just as plants are such products. Metallic minerals, like plants, were believed to be born and to grow inside the earth, from which they are harvested (mined). The first and largest fruits of unsown plants were selected each year and designated *mamas,* which suggests that they, along with unique metallic mineral specimens, were identified and named according to a fundamental Andean system of organization (Berthelot 1978; Salazar-Soler 1992).

In their myths and in ritual practice, Julcani miners today preserve the continuity of these traditions. A myth that recounts the origin of metals describes their birth in the order gold, silver, copper, lead. Julcani miners believe that of all the metallic minerals, only gold has completed its cycle of growth and maturation inside the earth; thus gold is the sole metal, the *mama,* to have been born. The other metals—silver, copper, lead, mercury—are incomplete, and miners consider them to be fetuses or stillborn (Salazar-Soler 1992).

Urton (1997) comments that the classification of metals, by both these contemporary miners and by Andean peoples in prehistory, follows a fundamental paradigm that operates in the Quechua language. Urton shows that nonnumeral, ordinal-like sequences—such as the fingers of a hand, the tines on a pitchfork, the vertical position of a sequence of ears of corn on a stalk, the interrelated groups of *ayllus* (kin groups) occupying a given territory, the colors of the rainbow—are conceptualized in Quechua in terms of reproductive processes and kinship relations: "Ordination is constructed as a set of relationships having as its prototype a 'natural' set of four or five members beginning in a principle associated with *mama,* the mature, reproductive female, and ending in an immature, essentially nongendered member of a consanguineal group, the child" (Urton 1997: 87). Urton argues that the role of *mama* as "progenitrix" in Quechua systems of classification operates for metals as for other sets that order members through states or stages of reproduction. Metals grow inside the earth from a common *mama,* gold; silver, copper, lead, and mercury follow in sequence, as less developed, "weaker" members of the line.

This metal set is unlike the other ordinal-like sequences Urton uses as examples of Quechua classificatory schemes, however, for in all of those some physical relationship associates the members: fingers to hand, tines to pitchfork, color bands to rainbow, corn cobs to corn stalk, *ayllus* to common territory. The link between *mama* and offspring is readily apparent. In contrast, the metal sequence is founded on no such physical associations. The physical features that assemble metallic minerals in a natural group are their location within the earth, and the need for people to intervene and excavate them from a subterranean environment in order to process them for their metallic content. A mine exists only when miners elect to exploit an ore; under such circumstances, all metallic ores are physically associated with mines. But the ores are not necessarily associated with one another—as are fingers or tines. In a polymetallic deposit, certain ores (such as those of silver, lead, and zinc) may be present in zoned association. But others in the classificatory set will not be present. Some deposits are rich in a single mineral species, yielding none of the other members of the set. Finally, metallic ores of one species do not alter to ores of another. There are no physical or chemical processes by which such transmutation occurs. It may be that in the case of metals, the analogy with plants that grow and mature to produce ripe fruit, encouraged the grouping of metallic minerals in a sequence of members with a "head": a *mama.* But the order among members likely represents cultural hierarchies established for them early in Andean prehistory and later preempted by the Inka. That order reflects social status, religious, and gender ascriptions that governed how and by whom these metals were used (Alva and Donnan 1993; Lechtman 1997).

The utility of a materials-archaeological approach in this instance, of focusing on what and how people *do* rather than on what and how they say, is that it confines us to detailed scrutiny of metals and their relationships in practice. Ethnocategories arise from patterns of technological practice, whether or not those patterns are labeled linguistically. My estimation that copper was the mother of metals among Andean smiths comes from the material facts of how copper was managed in association with other metals. It does not assign to copper the role of *mama,* or any other role, for that matter. Role assignment is a cultural function. It simply reports kinship between copper and other metals when they are alloyed together. In Andean metallurgy, copper was the metal whose presence enabled previous properties to be altered and new ones to be generated.

REFERENCES CITED

Allen, C. J.

1988 *The Hold Life Has: Coca and Cultural Identity in an Andean Community.* Smithsonian Institution Press, Washington, D.C.

Alva, W. L., and C. B. Donnan

1993 *Royal Tombs of Sipán.* Fowler Museum of Cultural History, University of California, Los Angeles.

Berthelot, J.

1978 L'exploitation des métaux précieux au temps des Incas. *Annales: Economies, Sociétés, Civilisations* 33(5–6):948–966.

Carpenter, L. K.

1992 Inside/Outside, Which Side Counts? In *Andean Cosmologies through Time,* edited by R. V. H. Dover, K. E. Seibold, and J. H. McDowell, pp. 115–136. Indiana University Press, Bloomington.

Conklin, W. J.

1996 Structure as Meaning in Ancient Andean Textiles. In *Andean Art at Dumbarton Oaks,* edited by E. H. Boone, pp. 321–328. Dumbarton Oaks Research Library and Collection, Washington, D.C.

Harrison, R.

1982 Modes of Discourse: The *Relación de antiguedades deste reyno del Piru* by Joan de Santacruz Pachacuti Yamqui. In *From Oral to Written Expression: Native American Chronicles of the Early Colonial Period,* edited by R. Adorno, pp. 65–99. Maxwell School of Citizenship and Public Affairs, Syracuse University, Syracuse.

Hosler, D.

1994 *The Sounds and Colors of Power: The Metallurgy of Ancient West Mexico.* MIT Press, Cambridge, Mass.

Hosler, D., H. Lechtman, and O. Holm

1990 *Axe-monies and Their Relatives.* Studies in Pre-Columbian Art and Archaeology No. 30. Dumbarton Oaks, Washington, D.C.

Lechtman, H.

1973 The Gilding of Metals in Pre-Columbian Peru. In *Application of Science in Examination of Works of Art,* edited by W. J. Young, pp. 38–52. Museum of Fine Arts, Boston.

1977 Style in Technology: Some Early Thoughts. In *Material Culture: Styles, Organization, and Dynamics of Technology,* edited by H. Lechtman and R. S. Merrill, pp. 3–20. West Publishing Co., St. Paul, Minn.

1984 PreColumbian Surface Metallurgy. *Scientific American* 250:56–63.

1988 Traditions and Styles in Central Andean Metalworking. In *The Beginning of the Use of Metals and Alloys,* edited by R. Maddin, pp. 344–378. MIT Press, Cambridge, Mass.

1993 Technologies of Power: The Andean Case. In *Configurations of Power in Complex Society,* edited by J. S. Henderson and P. J. Netherly, pp. 244–280. Cornell University Press, Ithaca, N.Y.

1996a Cloth and Metal: The Culture of Technology. In *Andean Art at Dumbarton Oaks,* edited by E. H. Boone, pp. 33–43. Dumbarton Oaks Research Library and Collection, Washington, D.C.

1996b Andean Metallurgy. In *Pre-Columbian Science,* edited by A. F. Aveni. *Storia della Scienza.* Istituto della Enciclopedia Italiana, Rome (in press).

1997 The Inka, and Andean Metallurgical Tradition. In *Variations in the Expression of Inka Power,* edited by R. Matos, R. Burger, and C. Morris. Dumbarton Oaks, Washington, D.C. (in press).

Money, M.

1998 *El Significado de los metales en los Andes de acuerdo a los diccionarios Aymara y Quechua del Siglo XVII.* Unpublished Ph.D. dissertation, Department of History, Columbia University, New York.

Salazar-Soler, C.

1992 Magia y modernidad en las minas andinas. Mitos sobre el origen de los metales y el trabajo minero. In *Tradición y modernidad en los Andes,* edited by H. Urbano, pp. 197–219. Centro de Estudios Regionales Andinos "Bartholemé de las Casas," Cusco.

Smith, C. S. S.

1975 Metallurgy as a Human Experience. *Metallurgical Transactions* A6A(4):603–623.

Taylor, G.

1974–1976 *Camay, camac,* et *camasca* dans le manuscrit Quechua de Huarochirí. *Journal de la Société des Américanistes* 63:231–244.

Urton, G. D.

1981 *At the Crossroads of the Earth and the Sky: An Andean Cosmology.* University of Texas Press, Austin.

1997 *The Social Life of Numbers: A Quechua Ontology of Numbers and Philosophy of Arithmetic.* University of Texas Press, Austin.

Index

Page numbers in italics indicate figures